About revision

INTRODUCTION

If anyone has discovered an instant way to guarantee success in exams, they are keeping very quiet about it! This book does not provide any magic answers, either, but it should help you find a method of revising for exams which will work for *you*. Developing a good, workable revision technique is very important. As you progress through school, and maybe on to some form of further education, you will find you have to cope with increasingly large quantities of information which you will need to understand, learn and apply. Acquiring the skill and confidence to cope with all this information is a very important part of learning.

One aspect of revision involves finding out about your strengths, and building on them. To find out what sort of a student you are, have a go at the quiz below. Be honest!

Studying for exams

1 How long do you usually study for in the evening? **(a)** around two hours **(b)** as long as is necessary to get everything done **(c)** as short a time as possible	**6** How far ahead do you generally plan your work? **(a)** a few weeks ahead, if possible **(b)** a few days at the most **(c)** don't really plan ahead
2 How easy do you find it to get started on your work in the evening? **(a)** no problem **(b)** quite hard **(c)** very difficult	**7** How much do you worry about exams? **(a)** quite a lot **(b)** almost all the time **(c)** try not to think about them
3 How easy do you find it to concentrate on your work once you get started? **(a)** no problem **(b)** usually fairly hard **(c)** often distracted	**8** When do you start your revision for exams? **(a)** a couple of months before the exams start **(b)** the week before the exams start **(c)** the night before the exams start
4 Where do you usually study? **(a)** in a quiet room **(b)** in a room used by other people **(c)** on the bus going to school	**9** How well do you understand your notes when revising for exams? **(a)** everything understood **(b)** most things understood **(c)** can't always find your notes
5 How do you get most of your homework done? **(a)** to a planned timetable, if possible **(b)** do it the night it is set **(c)** do it at school in the morning and in breaks	**10** What do you do when you are revising for exams? **(a)** summarize notes, test yourself, try questions **(b)** read through notes **(c)** give up – there's just too much to cope with

Now turn the page to find out what sort of student you are...

Award yourself points on the following scale:
(a) answers = 5 points   (b) answers = 3 points   (c) answers = 1 point

50 POINTS

Congratulations! You are a super student – or you are good at getting maximum points on quizzes like this! If you really were honest, then you are well on your way to getting good exam results. Read on to find out how this book will help you.

20-49 POINTS

Congratulations! You are a real student. You are organized up to a point, and generally get most of your work done. You probably worry quite a lot about exams, and you often feel your efforts are not rewarded in your grades. You are probably the sort of person who takes your books home well before the exams start, but you never really seem to get down to revision. You have probably had 'night before the exam' panics when you open your books for the first time and realise how much you have *not* done! Read on to find out how this book will help you.

19 POINTS OR LESS

The fact that you've got so far means that you care about your work and you'd like to do as well as possible in exams. You may also have been too hard on yourself and your study methods when you answered the quiz. What you find hard is actually getting started and keeping going. Read on to find out how this book will help you.

How this book will help you

Whatever sort of student you are, this book will help you do as well as you possibly can in your Salters' Science GCSE exam. This is because the tasks in this book have been drawn together with two key points in mind:

- they will require you to revise *actively*, and so make your revision as effective as possible and help you remember what you have done;

- they are *brief*, so that you can be flexible about how you fit them into your own personal revision schedule for all your subjects.

A few brief facts about the exam

All your work over the past two years is leading to an exam whose full title is *The Sciences: Double Award (Salters' Science)*.

Your final mark will be based on:

- the assessed practical work you have done throughout the course (20%);
- your two literature-related coursework assignments (10%);
- the two or three examination papers you sit at the end of the course (70%).

Papers 1 and 2 are each 1 hour 30 minutes long, and consist of short-answer questions. The optional Paper 3 is 2 hours 15 minutes long, and contains some short-answer questions and some longer questions. All the questions on each paper are compulsory.

You will find out more about the different sorts of exam questions and how to tackle them in the activities in this book.

Did you know ... ?

An average of one million chemical reactions take place in your brain every minute.

Most people work best early on in the day or early on in the evening.

The ability to concentrate hard drops off after an hour of study. Short breaks at sensible stopping points can work wonders.

If you had lived 500 years ago, you would have thought your brain was near your heart, not in your head.

Exams are not just about what you have learned, but also knowing how to tackle the questions on the paper.

You are more likely to do well in exams if you work to an organised timetable – but a timetable that is realistic and flexible.

Most people revise best in a quiet, comfortable room with paper, pens, pencils and books close to hand.

Your revision will be most effective if it is as <u>active</u> as possible – mentally active at least! Active revision involves <u>hard thinking</u>, not just reading through notes or books. When you think about what you are doing, revision is more interesting AND you remember more.

Revising science ideas

ABOUT THIS BOOK

This book will help you revise for your GCSE Science examination. The Salters' Science course you have been following is unique because it sets scientific ideas (concepts) in particular situations (contexts). So, for example, you have met ideas about the electromagnetic spectrum in the units on COMMUNICATING INFORMATION and SEEING INSIDE THE BODY. This book will help you to link together all the important science concepts you have met during the course.

Parts of the book
To help you link together all these ideas, there are two main parts in the book. The first part of the book is divided into thirteen **themes.** The second part of the book contains the **unit summaries**, or *In briefs,* for each of the twenty-two Salters' units you have studied in your GCSE course. You may have seen these before if you have used the Salters' textbooks.

You may be wondering why thirteen themes have been chosen when you studied twenty-two units! This is because each of the themes concentrates on one or two of the big ideas that run through science. Often these big ideas appear in more than one unit. Sometimes people use slightly different words to describe the big ideas, but the science is just the same.

The themes
The thirteen themes used in this book to link together the big ideas are:
1 PROCESSES OF LIFE
2 THE EARTH AND ITS SURROUNDINGS
3 MATERIALS AND HOW WE USE THEM
4 FORCES AND HOW WE USE THEM
5 ENVIRONMENTS AND HOW WE AFFECT THEM
6 RADIATIONS AND HOW WE USE THEM
7 HOW MATERIALS CAN BE CHANGED
8 ENERGY RESOURCES AND TRANSFERS
9 NATURAL CYCLES AND HOW WE AFFECT THEM
10 EARTH AND UNIVERSE
11 SIMILARITIES AND DIFFERENCES IN LIVING THINGS
12 USING ELECTRICITY AND MAGNETISM
13 EXPLAINING HOW MATERIALS BEHAVE

The concept map
The **concept map** on the opposite page shows you where the thirteen big ideas appear within the UNIT GUIDES in the GCSE course.

If you follow the arrows linking the units on the concept map, you will see, for example, that ideas about the theme FORCES AND HOW WE USE THEM appear in the units MOVING ON, SPORTS SCIENCE and THE EARTH IN SPACE. Ideas about the theme THE EARTH AND ITS SURROUNDINGS are covered in the units RESTLESS EARTH and THE ATMOSPHERE.

Sections within themes
Each theme has three sections:

● *What do you know about ... ?* These are quick questions to get your brain 'tuned in' to the theme – and show you how much you already know about ideas covered in the theme!

● *Getting to grips with ...* These are activities for you to carry out using the key ideas in the theme. Carrying out these activities forms a vital part of your revision programme because you will be thinking hard about what you are doing. This will help you to understand and remember more about the key ideas in the theme.

● *Questions about ...* These are questions to help you assess your progress. In some cases, these will be exam-style questions, with tips on how to produce the best answers.

Using this book
You will benefit most from this book if you work through the themes in the order they appear, particularly in the case of the earlier themes. This is because this book is intended to do more than summarize the key science ideas you have met. It is also intended to help you develop *revision techniques which work for you* as you revise.

A concept map showing where the thirteen themes appear in the units.

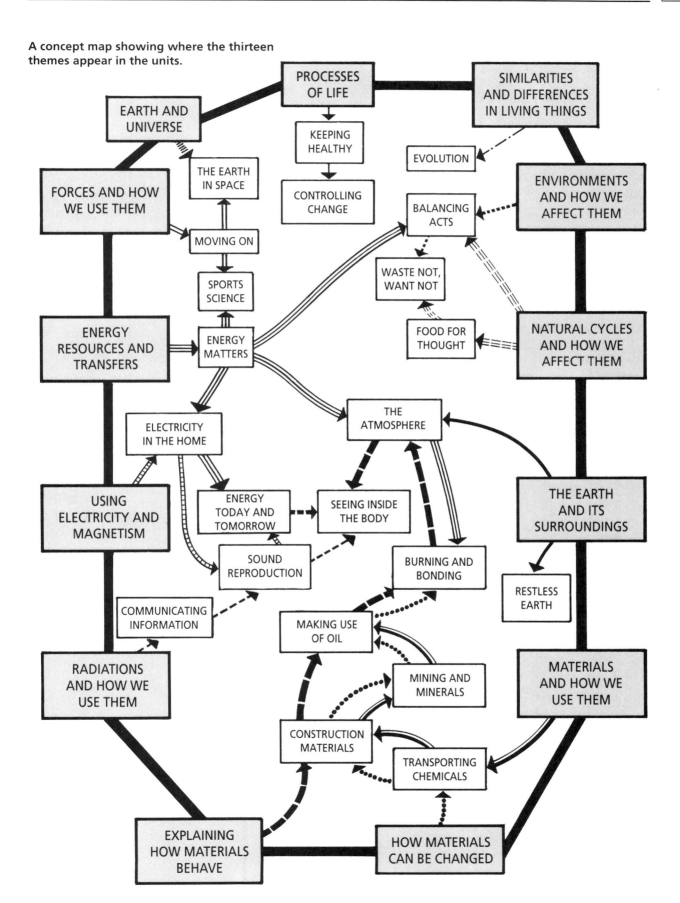

THEME 1 PROCESSES OF LIFE

To revise this theme, you will need to refer to your notes and the *In briefs* for the units KEEPING HEALTHY and CONTROLLING CHANGE.

What do you know about PROCESSES OF LIFE?

Try the following questions. Check your answers by referring to the *In briefs* for the two units.

1 Someone you know is wondering why they had to have a urine test as part of a medical examination. Explain to them why a urine test is carried out.

2 Your local surgery offers injections against a flu virus. What is a virus?

3 You quite often hear on the news about organ transplants being carried out. Which body organs can be transplanted?

4 One of the two systems for controlling change in your body is your nervous system. What is the other system called?

5 What are the biological catalysts in your body called?

Getting to grips with PROCESSES OF LIFE

Activity 1

One way of beginning to organise your revision notes is to imagine you had to write a heading for each of the ideas in the *In briefs*. Try doing this for KEEPING HEALTHY and CONTROLLING CHANGE. For example, you might have **'Being ill'** for the first section in KEEPING HEALTHY, or **'Stimulus and response'** for the first section of CONTROLLING CHANGE. When you do this for the first time it is very useful to work with a friend, so that you can compare your ideas.

Make a note of your headings for each of the sections in the *In briefs*. You *could* just make a list of all the headings. However, you might instead want to put each heading onto a small piece of paper or card – this would make Activities 2 and 3 easier for you to carry out.

KEEPING HEALTHY
being ill
microbes
Killing microbes
germicides
clots
the immune system
vaccination
AIDS
transplants
kidneys
heart
enzymes
how enzymes work

CONTROLLING CHANGE
stimulus and response
change
systems for controlling change
hormones and nerves
tropisms
phototropism
enzymes
enzymes and feedback
enzymes and digestion
homeostasis
mammals and homeostasis
transpiration

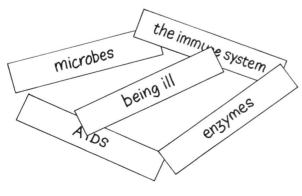

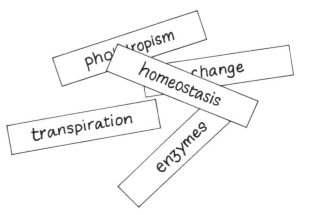

Activity 2

The next step is to look for links and connections between your headings, so you can group together the sections in the *In briefs* which have a common theme.

Look carefully at your list of headings. Are any of the headings *within* a unit linked? If so, group them together. You can do this very easily if you have your headings on pieces of paper or card.

If you have a list, you should draw a circle round headings which form a group of linked ideas.

Now look at your list of headings again. Are there links *across* units for any of your headings? If so, either group these cards together or use arrows on your list to show that the headings are linked.

Now try and think of headings which cover each of these linked ideas. Write these headings onto further pieces of card, or onto your list.

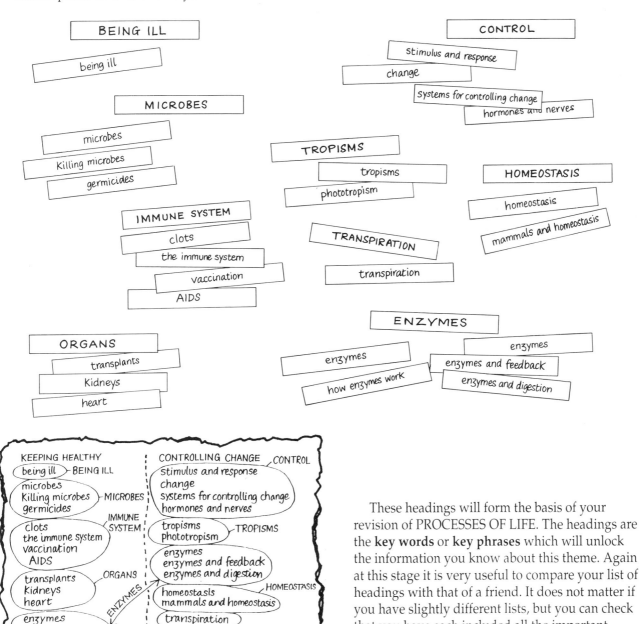

These headings will form the basis of your revision of PROCESSES OF LIFE. The headings are the **key words** or **key phrases** which will unlock the information you know about this theme. Again at this stage it is very useful to compare your list of headings with that of a friend. It does not matter if you have slightly different lists, but you can check that you have each included all the important ideas.

Activity 3

Once you have your list of key words, your next task is to make a summary of the main ideas which relate to each of the words.

The diagram below shows one way of making a summary of important ideas. This diagram is about microbes. The diagram is shown in handwriting. The notes in print show you how it was constructed.

This type of diagram is called a **spray diagram,** or sometimes a **spider plan.** Many people remember information in this form much better than information in paragraphs of text – you may be one of them!

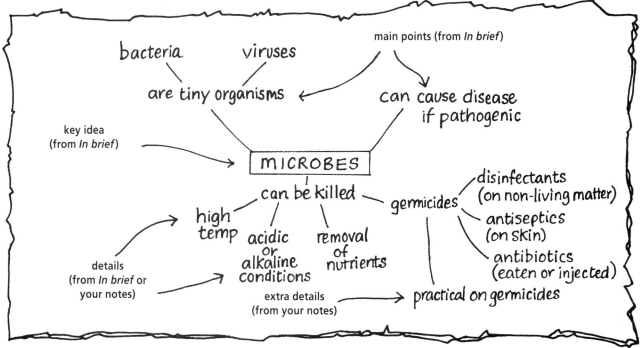

A spray diagram should be constructed with 'the four Cs' in mind. It should be:

● Clear – you need to be able to read it!

● Concise – don't put too much on it.

● Colourful – to make it more memorable.

● Comprehensive – you should have got all the big ideas in.

You can use your key word and key phrase headings to construct your own spray diagram summaries.

Now use the *In briefs* for KEEPING HEALTHY, together with your notes on the unit, to make a spray diagram for 'Becoming immune'. Your card headings will provide you with the key ideas and the main points. You can use the other information in the *In brief,* and refer to your notes to add extra details.

The questions below should help you make your summary:

1 In what ways can you become immune to microbes?
2 How does your immune system work?
3 Which diseases have been eliminated by vaccination?

Because it takes practice to produce good spray diagrams, you may well need to have a couple of goes before you are happy with the result. Depending on the size of your writing, and how neat you are, your final version could be on file paper, or on a postcard. It's also a good idea to compare your spray diagram with that of a friend so that you get an idea of how someone else has summarized the information. Don't worry if the summaries *look* different – the important thing is that all the key ideas should be there.

Activity 4

The idea is to build up a summary of all the main ideas. These summaries should contain more detail than the *In briefs*, but less than your notes. They should also concentrate on the scientific ideas. In producing the summaries, you have to think hard about what you are doing. This means you remember the information much better than you would if you just read through your notes.

Have a go at a spray diagram on 'Transplants'. Start by making a list of questions/points you want to include. You will find it very helpful to compare your list with that of a friend before you produce your diagram.

Activity 5

If producing all these summaries seems like a lot of work, *this* activity should show you why the effort is worth it. Try it about a week after you made your summaries.

Test yourself by covering up all the details and seeing how much you remember about each of the key words. You should get a pleasant surprise! And the more often you go through the summaries, the more you will remember.

Keep your summaries in a safe place! You will be adding to them as you work your way through each of the themes in this book.

Questions on PROCESSES OF LIFE

The real test of how effective your revision has been is how well you can answer questions on the topic. You might want to use the questions to test yourself, or you could ask a friend to try and answer the questions. If you know how well your friend is answering the questions, you must also know quite a lot yourself!

Start by going back to any of those questions at the start of the theme that you could not answer. Then have a go at the questions below, and see how much you have remembered …

To revise this theme, you will need to refer to your notes and the *In briefs* for the units RESTLESS EARTH and THE ATMOSPHERE.

What do you know about THE EARTH AND ITS SURROUNDINGS?

Try the following questions. Check your answers by referring to the *In briefs* for the two units.

1 Imagine you have found a sample of rock on a field trip. How would you try to find out what type of rock it was? What are the three main types of rock?

2 *Plate Movement Causes Earthquake.* What does this headline mean?

3 What is the inner layer of the atmosphere that surrounds us on Earth called?

4 How much of the air we breathe is oxygen? What other gases are there in air?

5 If you heard a weather forecaster talking about 'closely-packed isobars', what sort of weather would you expect? What are *isobars*?

Getting to grips with THE EARTH AND ITS SURROUNDINGS

Activity 1
Start by thinking of a heading for each of the sections in the *In briefs* for RESTLESS EARTH and THE ATMOSPHERE. For example, you might have 'Structure of the Earth' for the first section or 'Plates' for the third section in RESTLESS EARTH.

Now link some of your headings together to create the key words or key phrases for your summaries.

Below is one way of linking the sections in the *In briefs* together.

However, you may decide that you want to summarise the *In briefs* in a different way. For example, you might think that there is a lot to say about the weather, and you would prefer to have this in two or three sections. It's up to you! There is no one correct way of producing the key words or phrases. *You* will come up with the list which means most to *you*. It is useful, though, to compare your list with the list a friend has drawn up. That way you can check to see that you have each covered all the important ideas.

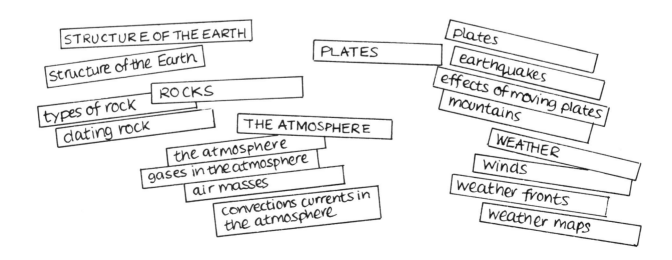

Activity 2

Now you need to make your summaries for each of your key words or phrases.

You might be happy doing sprays for your summaries. However, different types of summary work best for different people, and you might like to try a different kind of summary. It doesn't really matter what sort of summary you use so long as you find a style that works for you.

The important thing about making any kind of summary is that you have to think carefully about the information in order to decide what is important. *This* is what helps you to understand and remember!

The summary below has listed points under headings for the 'Structure of the Earth'.

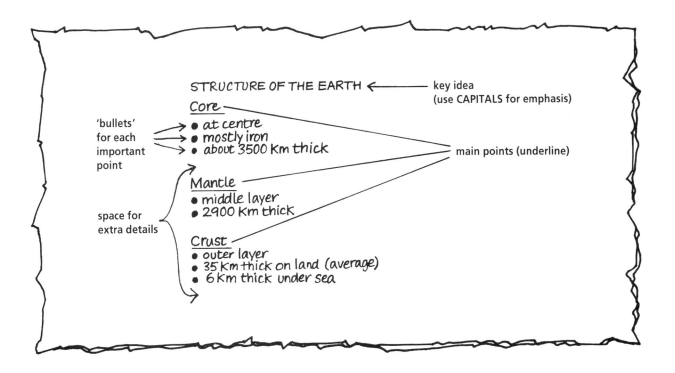

Activity 3

Don't forget to use your summaries to test yourself about a week later.

You should also go back and see how much you can remember about PROCESSES OF LIFE.

Questions on THE EARTH AND ITS SURROUNDINGS

Try these questions out on yourself or a friend.

Could you ...

- describe the different layers in the Earth?
- give some examples of the different types of rocks in the Earth?
- explain how earthquakes occur?
- explain why the remains of small sea creatures can be found at the tops of some mountains?
- say which gases are present in the atmosphere?
- describe some of the measurements meteorologists need to make to help them forecast the weather?

THEME 3 MATERIALS AND HOW WE USE THEM

This theme runs through several units. This is because materials are so important! To revise this theme, you will need to refer to your notes and the *In briefs* for the units TRANSPORTING CHEMICALS, CONSTRUCTION MATERIALS, MINING AND MINERALS and MAKING USE OF OIL. These units also cover ideas related to the themes HOW MATERIALS CAN BE CHANGED and EXPLAINING HOW MATERIALS BEHAVE.

What do you know about MATERIALS AND HOW WE USE THEM?

Try the following questions. Check your answers by referring to the *In briefs* for the units.

1 There are many different kinds of useful substances. Substances which cannot be broken down into anything simpler are called ....

2 Many objects can be made from plastics. Some plastics can be reshaped when they are heated, and these are called ...... plastics.

3 Limestone is a mineral which can be used as a material for building. It contains almost pure ....

4 Very reactive metals are extracted from minerals containing them by a process called ....

5 Crude oil is a mixture of compounds which have many uses. These contain mainly the elements carbon and .... These compounds are called ....

Getting to grips with MATERIALS AND HOW WE USE THEM

Activity 1

This activity will help you pick out the parts of the *In briefs* which are important for this theme.

Start with the *In brief* for TRANSPORTING CHEMICALS. Use the questions below to help you identify the parts of the *In brief* which concern MATERIALS AND HOW WE USE THEM.

- Why do we need a chemical industry?
- Where are chemical industries situated?
- Why do chemicals sometimes need to be transported?
- What safety precautions are needed when chemicals are transported?
- What is the difference between elements and compounds? How can their names be written in chemical shorthand?
- Why is the periodic table useful?

Now look at the *In brief* for CONSTRUCTION MATERIALS.

- What different types of materials are there? What are they used for? (Don't worry at the moment about the reasons why they behave the way they do – this comes later.)

Now go on to the *In brief* for MINING AND MINERALS.

- What are minerals?
- Where do they come from?
- Will they last for ever?
- How are they found?
- What decides whether or not a mineral will be extracted?
- What might be the benefits and drawbacks of a mine?

Finally, look at the *In brief* for MAKING USE OF OIL.

- Where is oil found?
- What does it consist of?
- What can it be used for?
- Will it last for ever?

You may want to compare the list of sections you have identified with the list a friend has produced.

Activity 2
Now you have identified the relevant sections of the *In briefs,* you can decide on your headings, and how you are going to group your headings together. Some ideas for key words and phrases are shown opposite:

THE CHEMICAL INDUSTRY

CHEMICAL SHORTHAND

TYPES OF MATERIALS

MINERALS

DECISIONS ABOUT MINING

OIL

Activity 3
By now you should be getting quite good at making summaries, and you should also have an idea of the sort of summary which works best for you.

Use the appropriate parts of the *In briefs* and your notes to make summaries of the key ideas in this theme.

Don't forget to test yourself a week later!

Questions on MATERIALS AND HOW WE USE THEM

The questions below are the sorts of questions you will find at the start of Papers 1 and 2 in the exam. They are questions which require either a letter or a one-word answer.

The materials listed below are used in building.

A concrete
B glass
C iron
D lime

Which of the materials A to D,
1   is made in a blast furnace?
2   uses soda ash in its manufacture?
3   is used to make mortar?
4   contains cement?

A aluminium
B bauxite
C aluminium oxide
D oxygen

From the list of substances, A to D, choose one which is
5   a pure compound.
6   a non-metallic element.
7   an ore.

A SO_2
B $CaCO_3$
C C_2H_6O
D Cl_2

From the formulas, A to D, choose the one which
8   contains only one element.
9   contains the element sulphur.
10   contains the element calcium.
11   represents the greatest number of atoms.

12   The formula NaOH is put on a tanker containing ....

13   The formula H_2SO_4 is put on a tanker containing ....

14   Ammonia is manufactured from nitrogen and ....

THEME 4 FORCES AND HOW WE USE THEM

To revise this theme, you will need to refer to your notes and the *In briefs* for the units MOVING ON, SPORTS SCIENCE and THE EARTH IN SPACE.

What do you know about FORCES AND HOW WE USE THEM?

Try the following questions. Check your answers by referring to the *In briefs* for the units.

1 Which of the following needs a force?

(a) getting a bicycle moving;
(b) stopping a moving bicycle;
(c) making a bicycle move faster;
(d) making a bicycle slow down;
(e) changing the direction a bicycle is moving in;
(f) keeping a bicycle moving in a straight line at a steady speed on a smooth flat surface.

2 How could you work out the driving force from a car's engine if you knew the mass of the car and how quickly it could accelerate away from traffic lights?

3 How could a safety helmet help protect your head if you had an accident when cycling?

4 A nutcracker is an example of a lever. It works by helping you to multiply the ...... you apply to the nut to crack the shell.

5 The force which keeps the Moon and satellites in orbit around the Earth is called the Earth's ...... force.

Getting to grips with FORCES AND HOW WE USE THEM

All the ideas in the *In brief* for MOVING ON are about FORCES AND HOW WE USE THEM. Ideas about forces also appear in Sections 11–16 of the *In brief* for SPORTS SCIENCE. More ideas about forces are covered in Sections 6–16 of the *In brief* for THE EARTH IN SPACE.

Start by deciding on headings for the relevant sections of the *In briefs*. Because there are lots of ideas in these units, and some are more difficult than others, you might find it particularly helpful to put these headings onto small pieces of card. This will help you when you come to group your headings together.

Now group your headings together. Some suggestions are given opposite.

In this particular theme, you might also want other summaries, for example:

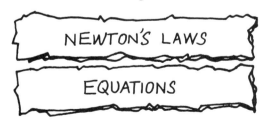

NEWTON'S LAWS

EQUATIONS

You will find that some exam questions provide you with the equations you need, but sometimes you are expected to remember which equation you need to use. So it's worth making the effort to learn equations – and also to learn when to use them!

Now you can draw up your summaries.

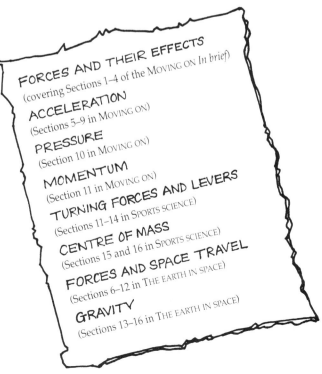

FORCES AND THEIR EFFECTS
(covering Sections 1–4 of the MOVING ON In brief)

ACCELERATION
(Sections 5–9 in MOVING ON)

PRESSURE
(Section 10 in MOVING ON)

MOMENTUM
(Section 11 in MOVING ON)

TURNING FORCES AND LEVERS
(Sections 11–14 in SPORTS SCIENCE)

CENTRE OF MASS
(Sections 15 and 16 in SPORTS SCIENCE)

FORCES AND SPACE TRAVEL
(Sections 6–12 in THE EARTH IN SPACE)

GRAVITY
(Sections 13–16 in THE EARTH IN SPACE)

Questions on FORCES AND HOW WE USE THEM

There are are exam-style questions at the end of this section.

Exams are not just about how much you know, they are also about exam technique.

- **R**eading the instructions carefully.
- **E**stimating how much time you have for each question.
- **A**nswering the question that is being asked.
- **D**eciding which questions you can get most marks on, as these are the ones to do first.

READ the initial letters to find the key word for exam technique!

You will find questions like the ones below on Section B of Papers 1 and 2. They are called **structured questions,** because they contain spaces where you write your answers.

Before you try the questions, you need to read the instructions:

Section B
Questions **B21–B26** Answer **all** questions.
Write your answers in the spaces provided.
Spend about 20 minutes on this section.

1. What are the general instructions?

Read the instructions.
E
A
D

4. How well can I answer the question?

R
E
A
Decide how much of the question you can do. Complete the questions you can do best first in case you run out of time. Remember also you don't lose marks for wrong answers, so it's always worth a sensible guess.

2. How much time do I have?

R
Estimate how much time you
A have for each question.
D (6 questions in 20 mins = about 3 mins per question + a couple of minutes to check at the end)

3. What's the question asking?

R
E
Answer the question being
D asked (Look at the amount of space and the marks for each part)

An athlete is training for a javelin competition. Her position on the track as she runs up to throw the javelin has been marked every second.

Scale 1 cm = 5 m

(a) (i) What distance did the athlete run before throwing the javelin?
....(1)

(ii) How long did it take the athlete to cover this distance?
....(1)

(iii) Calculate the athlete's average speed during the run.
....(1)

(b) Between which points was the athlete running at a steady speed? Explain your answer.
....
....(2)

(c) The athlete throws the javelin with a force of 50 N. The javelin has a mass of 2 kg. Calculate the acceleration of the javelin as it leaves her hand.
....(1)

Look carefully at the information on the diagram. Don't forget to include units!

You need to remember that average speed = distance/time.

Calculations often use answers from previous sections.

Explain...(2) There will be one mark for stating the answer, and one for your explanation.

THEME 5 ENVIRONMENTS AND HOW WE AFFECT THEM

To revise this theme, you will need to refer to your notes and the *In briefs* for the units BALANCING ACTS and WASTE NOT, WANT NOT. These units also cover ideas related to the theme NATURAL CYCLES AND HOW WE AFFECT THEM.

What do you know about ENVIRONMENTS AND HOW WE AFFECT THEM?

Try the following questions. Check your answers by referring to the *In briefs* for the units.

1 What factors would influence the population of foxes in an environment?

2 Give two examples of household waste which could be recycled.

3 Why are people being encouraged to recycle household waste?

4 Some plastic bottles are now made of *biodegradable* plastic. What *is* a biodegradable plastic?

5 The tiny organisms which help decompose sewage into simple chemicals are called ....

Getting to grips with ENVIRONMENTS AND HOW WE AFFECT THEM

This theme deals with ideas about species, where they live, and the things which can affect them, including humans. In producing your summaries, you should be concentrating on the ideas in Sections 1–5 and 11 of the *In brief* for BALANCING ACTS and the ideas in Sections 1–10 of the *In brief* for WASTE NOT, WANT NOT.

Start by making your list of headings for each of the relevant sections in the *In briefs*.

Then link your headings together and decide on headings for each linked section.

Now compare your headings with those of a friend. Remember, this is to check that you have both got all the main ideas in, not that your headings are identical! One suggestion is shown opposite:

Now you can make your summaries.

> **FACTORS AFFECTING ENVIRONMENTS**
> (Section 5 in BALANCING ACTS)
> **EFFECTS OF HUMANS**
> (Sections 1–4 in BALANCING ACTS and Section 1 in WASTE NOT, WANT NOT)
> **LOOKING AFTER ENVIRONMENTS**
> (Section 11 in BALANCING ACTS)
> **TYPES OF WASTE**
> (Sections 1 and 5–7 in WASTE NOT, WANT NOT)
> **DEALING WITH WASTE**
> (Sections 2–4 and 8–10 in WASTE NOT, WANT NOT)

Questions on ENVIRONMENTS AND HOW WE AFFECT THEM

Here is another short question for you to try.

> 'Reuse, Recycle, Reclaim'
> The above slogan is used by an environmental group. In this year's campaign they are concentrating on encouraging people to collect empty drinks cans and old electricity cables and to return milk bottles.
> (a) For each of these campaigns, name the material which they hope will be saved.
> ....
> ....
> ....(3)
> ....
> (b) Explain which word from the slogan can be used most accurately in each of these campaigns.
> ....
> ....
> ....
> ....
> ....(6)

> Explain ... (6)
> You must say why you have chosen each word. There will be one mark for each correct choice and one for each correct explanation.

To revise this theme, you will need to refer to your notes and the *In brief* for the units COMMUNICATING INFORMATION, SOUND REPRODUCTION and SEEING INSIDE THE BODY. These units also cover ideas in the themes USING ELECTRICITY AND MAGNETISM and EXPLAINING HOW MATERIALS BEHAVE.

What do you know about RADIATIONS AND HOW WE USE THEM?

Try the following questions. Check your answers by referring to the *In briefs* for the units.

1 An electric fire, a light bulb and a radio speaker all emit radiation when they are switched on. Radiation transfers ...... from one place to another.

2 Visible light is one of the radiations of the ...... spectrum.

3 Your local radio station broadcasts on a *wavelength* of 296 m and at a *frequency* of 101.4 MHz. What is the difference between the wavelength and the frequency of a wave?

4 You can hear someone speaking even if you cannot see them because sound can be reflected and diffracted. What is the difference between *reflection* and *diffraction* of radiations?

5 People are only treated with X-rays and radiations from radioactive materials in controlled amounts. These radiations are harmful because they can ...... materials as they pass through them.

Getting to grips with RADIATIONS AND HOW WE USE THEM

This theme covers all the ideas in the *In brief* for COMMUNICATING INFORMATION, Sections 1–13, 18 and 19 of SEEING INSIDE THE BODY; and Sections 1–9 of SOUND REPRODUCTION.

You might want to try a different approach to this section. Rather than make your own list of headings and grouping them together, use the list here as a starting point.

Which sections of the *In briefs* would you include under each key word or phrase? Are there any sections in the *In briefs* you would group together differently?

Now make your own summaries.

This theme provides you with a good opportunity to have a go at a **mnemonic,** or memory aid. Most people have heard of the mnemonic **R**ichard **O**f **Y**ork **G**ave **B**attle **I**n **V**ain, where the initial letters help you remember the colours in the spectrum of visible light. Mnemonics are particularly useful when you have to remember a number of things in a particular order. Often, the dafter they sound, the easier they are to remember!

See whether you can think of a mnemonic to help you remember all the radiations in the electromagnetic spectrum. Compare your mnemonic with those of your friends – who has thought of the most memorable?

You might want to go back to other themes and see whether there are ideas where a mnemonic would help. You could, for example, try to think up one for metals in their order of reactivity.

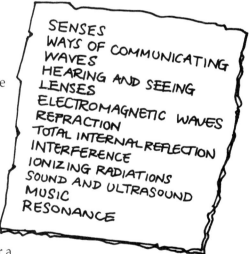

SENSES
WAYS OF COMMUNICATING
WAVES
HEARING AND SEEING
LENSES
ELECTROMAGNETIC WAVES
REFRACTION
TOTAL INTERNAL REFLECTION
INTERFERENCE
IONIZING RADIATIONS
SOUND AND ULTRASOUND
MUSIC
RESONANCE

Questions on RADIATIONS AND HOW WE USE THEM

Here are two more short structured questions for you to try.

This diagram shows the electromagnetic spectrum.

Radio   A   Visible   UV   B   Gamma

(a) State what types of radiations A and B represent.

....

....(2)

(b) (i) State which of the radiations is used for cooking food.

....(1)

(ii) State one other use for this type of radiation.

....(1)

(c) (i) Give one example of a radiation on the electromagnetic spectrum which can be used in medical diagnosis or treatment.

....(1)

(ii) Explain how the radiation is used.

....

....(2)

Many people in the world are short of food, but 25% of the world's food supply is lost after harvesting because it is not preserved adequately.

Gamma radiation can be used to preserve food. It can kill bacteria, delay the speed of ripening and sterilize sealed containers.

(a) Give two examples of processes which can lead to food decay.

....

....(2)

(b) Explain why the gamma radiation can sterilize food in sealed containers but not in open containers.

....

....(2)

(c) Some people are concerned about the possible effects of using an ionizing radiation on food because they think that the food may become radioactive.

(i) Explain what is meant by the term *ionizing* radiation.

....

....(2)

(ii) Would you expect food treated with gamma radiation to become radioactive? Explain your answer.

....

....(2)

To revise this theme, you will need to refer to your notes and the *In briefs* for the units TRANSPORTING CHEMICALS, CONSTRUCTION MATERIALS, MINING AND MINERALS, MAKING USE OF OIL and BURNING AND BONDING. These units also cover ideas related to the themes MATERIALS AND HOW WE USE THEM and EXPLAINING HOW MATERIALS BEHAVE.

What do you know about HOW MATERIALS CAN BE CHANGED?

Try the following questions. Check your answers by referring to the *In briefs* for the units.

1 What are the raw materials for making glass?

2 For rusting to take place, both ...... and ...... must be present.

3 Rocksalt is a very important mineral because it contains the compound .... The substances ...... and sodium ...... can be made from it.

4 Why is the process of cracking of crude oil so important?

5 Butane is a hydrocarbon fuel used as bottle gas. Complete the equation to show what happens when it burns in air:

$$C_4H_{10} + ?O_2 \rightarrow 4CO_2 + ?$$

Getting to grips with HOW MATERIALS CAN BE CHANGED

Activity 1
You need to start by picking out the sections in the *In briefs* which relate to this theme.
    The key to identifying these sections is that they will describe starting materials and how these are made into different products.

Activity 2
Once you have identified the sections, you can make your summaries.
    Here are some suggestions for key words and phrases. You may well have different ideas, and there are more to add to this list!

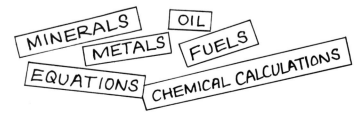

Activity 3

If you have followed the themes in this book in order, you are now more than half way through your revision. Not only that, you will also have developed your skills in making summaries of topics and done a lot of hard thinking about many of the important ideas in the course.

Keep using them to test yourself. Even a quick five-minute glance over one summary will help.

Every time you look at and think about a summary, you are transferring more information from your **short-term memory** to your **long-term memory.**

Your short-term memory is the part of your brain which will remember something like a new telephone number for as long as it takes to dial the number. Your long-term memory is the part of your brain that nearly always remembers things like telephone numbers you use frequently, without having to look them up.

The more you use your summaries, the more you will remember!

Questions on HOW MATERIALS CAN BE CHANGED

Here are two more short structured questions for you to try.

A company is investigating the recycling of tin from tin cans.

Tin cans are made of steel coated with a layer of tin.

(a) Why is the steel coated with a layer of tin?

....(1)

(b) The first problem in recycling the tin is to remove it from the steel. This can be done by reacting the can with hot hydrochloric acid.

tin + hydrochloric acid → tin chloride + hydrogen

Complete and balance the symbol equation for the reaction.

$$Sn + \ldots\ldots \rightarrow SnCl_2 + H_2$$

(2)

(c) Explain the safety problems which arise because

(i) hydrochloric acid is used in the process.

....
....(1)

(ii) hydrogen gas is produced in the process.

....
....(1)

A company which quarries limestone is planning to open a new quarry. They intend to use all the limestone to produce quicklime. This is done by heating the limestone. It decomposes to quicklime (calcium oxide) and carbon dioxide.

The company aim to produce 2800 tonnes of quicklime each week. They can find out how much limestone they will need to quarry by making use of this equation.

$$CaCO_3 \rightarrow CaO + CO_2$$

The relative atomic masses are:

Ca=40, O=16, C=12

(a) What is the relative molecular mass of

$CaCO_3$...

CaO...

CO_2...(3)

(b) What is the minimum mass of limestone which would need to be heated to produce 2800 tonnes of quicklime each week?

....
....(2)

Because energy is such an important idea in science, it crops up everywhere! However, you will be able to revise the main ideas in the theme if you refer to your notes and the *In briefs* for the units ENERGY MATTERS, SPORTS SCIENCE and ENERGY TODAY AND TOMORROW. Ideas about energy are also mentioned in THE ATMOSPHERE, BALANCING ACTS, ELECTRICITY IN THE HOME and BURNING AND BONDING.

What do you know about ENERGY RESOURCES AND TRANSFERS?

Try the following questions. Check your answers by referring to the *In briefs* for the units.

1 The total amount of energy after a food mixer has been used is the same as it was before. We say the energy has been ....

2 Fill in the gaps in this energy arrow diagram:

3 A food mixer has a power rating of 900 W. What information does this give you about the food mixer?

4 A child at the top of a slide has ...... energy. This is converted to ...... energy as she moves down the slide.

5 Why do people think it is important to conserve the Earth's energy resources?

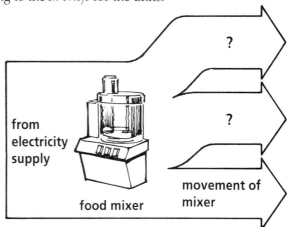

from electricity supply

food mixer

? ?

movement of mixer

Getting to grips with ENERGY RESOURCES AND TRANSFERS

Activity 1
The sections in the *In briefs* which relate to this theme can be found mainly in ENERGY MATTERS, SPORTS SCIENCE (Sections 1–10) and ENERGY TODAY AND TOMORROW (Sections 1–3 and 6–15).

   Start by deciding on headings for each of these sections. Then add to this list headings for the sections in the *In briefs* where reference is made to energy in the other units mentioned above.

Activity 2
Now group your headings together. Because there are many ways of grouping together all these ideas, this is another occasion where writing your headings on small pieces of card would be particularly helpful. You may well want to work with a friend when you come to group the ideas together.

Activity 3
Before you prepare your summaries, use the following questions as a guide to check that you have covered all the main ideas. Answers to these questions should also help you with ideas for your key words and phrases.

- What are the main sources of energy available to us?
- What forms can energy take?
- How can energy arrow diagrams be used to represent energy transfers?
- How does energy become spread out?
- How is energy supplied to your home?
- How can energy consumption in your home be measured?
- How can energy consumption in your home be reduced?
- How does energy transfer relate to work and power?
- How can electrical energy be generated on a large scale?
- What problems can be associated with large-scale generation of electrical energy?
- What are renewable energy resources?
- What energy transfers are associated with chemical reactions?

Questions on ENERGY RESOURCES AND TRANSFERS

The exam-style question at the end of this section is also an example of a **structured question.** You will find this type of question in Section C of Papers 1 and 2, and Section A of Paper 3. These questions carry more marks than the Section B-style questions, and generally require longer answers.

Don't forget the READ technique!

Now have a go at this type of question.

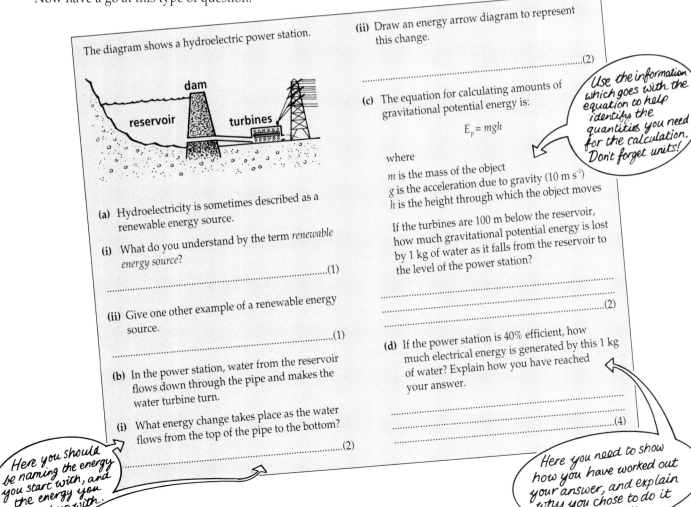

The diagram shows a hydroelectric power station.

(a) Hydroelectricity is sometimes described as a renewable energy source.

(i) What do you understand by the term *renewable energy source*? ....(1)

....

(ii) Give one other example of a renewable energy source. ....(1)

....

(b) In the power station, water from the reservoir flows down through the pipe and makes the water turbine turn.

(i) What energy change takes place as the water flows from the top of the pipe to the bottom? ....(2)

(ii) Draw an energy arrow diagram to represent this change.

....(2)

(c) The equation for calculating amounts of gravitational potential energy is:

$$E_p = mgh$$

where

m is the mass of the object
g is the acceleration due to gravity (10 m s^{-2})
h is the height through which the object moves

If the turbines are 100 m below the reservoir, how much gravitational potential energy is lost by 1 kg of water as it falls from the reservoir to the level of the power station?

....
....(2)
....

(d) If the power station is 40% efficient, how much electrical energy is generated by this 1 kg of water? Explain how you have reached your answer.

....
....
....(4)
....

To revise this theme, you will need to refer to your notes and the *In briefs* for the units BALANCING ACTS, FOOD FOR THOUGHT and WASTE NOT, WANT NOT. You have already come across some of the ideas in these units in the theme ENVIRONMENTS AND HOW WE AFFECT THEM.

What do you know about NATURAL CYCLES AND HOW WE AFFECT THEM?

Try the following questions. Check your answers by referring to the *In briefs* for the units.

1 The use of fertilizers is an important way of increasing food production. Name one problem which can be caused by their over use.

2 Give an example of one other method which can be used to improve world food production.

3 Gulls eat herrings. Animals which eat other animals are called .... The animals they eat are called .... These can be linked by a ...... chain.

4 *No artificial additives or preservatives.* Why are additives and preservatives sometimes put into foods?

5 Name two elements which are recycled naturally. How do humans influence these natural cycles?

Getting to grips with NATURAL CYCLES AND HOW WE AFFECT THEM

Start by identifying the relevant sections in the *In briefs*. Before you prepare your summaries, use the questions to help you with ideas for your key words and phrases.

- How has world food production been increased?
- What are the benefits and drawbacks of fertilizers?
- How can food be preserved and stored?
- What are food chains?
- How are carbon and nitrogen recycled naturally? How do humans interfere with these cycles?

Questions on NATURAL CYCLES AND HOW WE AFFECT THEM

Here is another longer structured question for you to try.

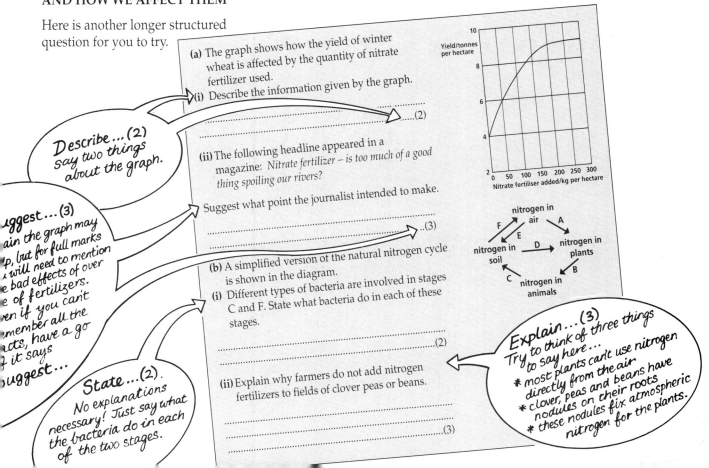

(a) The graph shows how the yield of winter wheat is affected by the quantity of nitrate fertilizer used.
(i) Describe the information given by the graph.
....
..............................(2)

(ii) The following headline appeared in a magazine: *Nitrate fertilizer – is too much of a good thing spoiling our rivers?*
Suggest what point the journalist intended to make.
....
....(3)

(b) A simplified version of the natural nitrogen cycle is shown in the diagram.
(i) Different types of bacteria are involved in stages C and F. State what bacteria do in each of these stages.
....
....(2)

(ii) Explain why farmers do not add nitrogen fertilizers to fields of clover peas or beans.
....
....
....(3)

Describe...(2)
say two things about the graph.

Suggest...(3)
...ain the graph may ...p, but for full marks ...will need to mention ...e bad effects of over ...e of fertilizers. ...en if you can't ...member all the ...cts, have a go ...f it says ...uggest...

State...(2).
No explanations necessary! Just say what the bacteria do in each of the two stages.

Explain...(3)
Try to think of three things to say here...
* most plants can't use nitrogen directly from the air
* clover, peas and beans have nodules on their roots
* these nodules fix atmospheric nitrogen for the plants.

To revise this theme, you will need to refer to your notes and the *In brief* for the unit THE EARTH IN SPACE. You have already come across some of the ideas in this unit in the theme FORCES AND HOW WE USE THEM.

What do you know about EARTH AND UNIVERSE?

Try the following questions. Check your answers by referring to the *In brief* for the unit.

1 Name the missing planets:

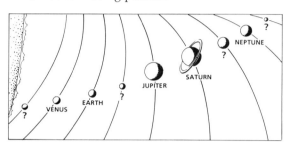

2 What other bodies can be found in the solar system?

3 How do the planets further from the Sun differ from those closer to the Sun?

4 What process takes place in the Sun to release energy?

5 What can happen to a star when its fuel runs out?

Getting to grips with EARTH AND UNIVERSE

This is a short topic, so you should have no problems making your summaries!

This might also be a useful time to go back and revise some of the more difficult ideas about forces which you met in THE EARTH IN SPACE.

Questions on EARTH AND UNIVERSE.

Try this longer structured question.

Below are data cards for three of the planets in the solar system.

PLANET A

Composition: Ice and rock
Orbital period: 84 years
Rotation period: 23 hours
Surface gravity: 1.3
Number of moons: 14
Average surface temperature: 60 K

PLANET B

Composition: Rocky
Orbital period: 224 days
Rotation period: 243 days
Surface gravity: 0.90
Number of moons: 0
Average surface temperature: 750 K

PLANET C

Composition: Mainly ice
Orbital period: 12 years
Rotation period: 10 hours
Surface gravity: 0.7
Number of moons: 17
Average surface temperature: 163 K

(a) Use the information on the cards to put the planets in order of increasing distance from the Sun. Explain how your arrived at your answer. (4)

(b) The Earth has an *orbital period* of 365 days and a *rotational period* of 24 hours.
(i) Explain what is meant by the two terms in italics. (2)
(ii) If the Earth's orbital period and rotational period were both longer, what effects would be noticed by people on Earth? Explain your answers. (4)

(c) Explain why the surface gravity on each of the planets is different. (2)

Here you need to look carefully for clues on the cards to help you reach an answer. You should be looking for patterns and trends in the data.

Here you need to think first about what someone on the Earth would notice as the Earth rotated and as it travelled round the Sun. This will help you identify the changes which would take place and help you explain the changes. There will be one mark for each effect, and one for each explanation.

To revise this theme, you will need to refer to your notes and the *In brief* for the unit EVOLUTION.

What do you know about SIMILARITIES AND DIFFERENCES IN LIVING THINGS?
Try the following questions. Check your answers by referring to the *In brief* for the unit.

1 Someone who has naturally red hair has ...... this characteristic from their parents. Someone who likes music has ...... this characteristic as they have grown up.

2 The theory which says that individuals with characteristics best suited to their environments are most likely to survive is called the theory of ....

3 Which part of a cell carries the coded information needed for control and development?

4 Chromosomes contain the chemical ...... , which is twisted into a double ...... shape.

5 Explain why it is possible for two children in a family to have blue eyes even though their parents both have brown eyes.

Getting to grips with SIMILARITIES AND DIFFERENCES IN LIVING THINGS
Make your summaries for the theme. Although this is a short theme, it covers some quite hard ideas, so take your time and make sure you understand the ideas.

Here are some suggestions for key words and phrases.

> DIFFERENT TYPES OF LIFE FORMS
> THEORIES OF ORIGINS OF LIFE FORMS
> CELLS AND INHERITANCE
> CHANGING LIFE FORMS

Questions on SIMILARITIES AND DIFFERENCES IN LIVING THINGS
The question on the next page is an example of the sort of question you will find in Section B of Paper 3. This type of question is called a **free response question.** This means there are no spaces within the question for your answers. Instead you have to use 'clues' in the question to help you structure your answer. This is the sort of question which requires the most thinking about before you put pen to paper.

The READ technique still applies, though!

'These new varieties of wheat have much shorter stalks than in my day', remarked the old farmer.

'Yes' said the grandson. 'These plant breeders are clever! The short stalks make it easier for us to harvest with a combine and it still gives a good yield.'

(a) Which two features of wheat plants did the plant breeders select as desirable characteristics? (2)

(b) Explain how a stock of seeds able to grow into plants with these characteristics might be produced. (2)

(c) Explain why short stalked wheat plants are unlikely to have evolved naturally. (4)

(d) One gene in wheat controls how soft the seeds are. The grandson counted that in 500 seeds, there were 132 soft seeds and 368 hard seeds.
Use your knowledge of inheritance to explain these numbers. (4)

Look for clues in the question to get you started here.

Explain...The numbers of marks for each of these sections suggest how many points you should mention in your explanation.

For example, in (c) there would be one mark for the following: ✳ideas about natural selection/survival of the fittest. ✳plants needing sunlight. ✳tall plants getting more sunlight. ✳tall plants surviving better.

To revise this theme, you will need to refer to your notes and the *In briefs* for the units ELECTRICITY IN THE HOME, SOUND REPRODUCTION and ENERGY TODAY AND TOMORROW. These units also have links with the theme ENERGY RESOURCES AND TRANSFERS.

What do you know about USING ELECTRICITY AND MAGNETISM?

Try the following questions. Check your answers by referring to the *In briefs* for the units.

1 This circuit contains a *cell*, a *resistor* and an *ammeter*. Which symbols represent each of these parts? What does the ammeter measure?

2 If the cell had a voltage of 1.5 volts, and you knew the reading on the ammeter, how could you work out the size of the resistance?

3 The lights in your house are connected together in a *parallel circuit*. What is a parallel circuit?

4 Sound is recorded on a tape using an *electromagnet*. Sound is played back from a tape by *electromagnetic induction*. What is the difference between electromagnetism and electromagnetic induction?

5 Telephone communication systems contain *transformers*. What does a transformer do?

Getting to grips with USING ELECTRICITY AND MAGNETISM

Start by identifying the sections in the *In briefs* which concern this theme. Once you have your headings for these sections, you can draw up your summaries.
   Here are some suggestions for key words and phrases:

> STATIC ELECTRICITY CURRENT TYPES OF CIRCUITS THE CIRCUIT EQUATION POWER CELLS
> THE NATIONAL GRID STORING AND REPRODUCING SOUND ELECTROMAGNETISM
> TRANSFORMERS ELECTROMAGNETIC INDUCTION

This is also a topic where the following summaries would be useful:

> EQUATIONS

> CALCULATIONS

Questions on USING ELECTRICITY AND MAGNETISM

Here is another example of a free-response question for you to try.

Here there will be one mark for each correct symbol.

Here you will need to say what actually happens, and explain why to get both marks.

Here you need to think about which equation you should use. The quantities mentioned in the question will give you clues.

Remember that calculations usually involve fairly easy numbers. This can often give you a clue as to what you need to do with the numbers.

The diagram shows how the fuel gauge on a car works. The sliding contact can move up and down the coil of resistance wire. When the fuel tank is empty, the contact is at X. When it is full, the contact is at Y.

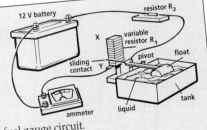

(a) The circuit diagram shows part of the fuel gauge circuit. Draw the completed circuit to show the ammeter and the resistance R_2. (2)

(b) Explain what happens to the ammeter reading as the tank is filled. (2)

(c) The resistance of the variable resistor R_1 can vary from 0 to 108 Ω. R_2 has a resistance of 12 Ω. The resistance of the ammeter is so small that it can be ignored. Calculate the ammeter reading when:
   (i) the tank is full (2)
   (ii) the tank is empty (2)

(d) Explain why the resistance R_2 is included in the circuit. (2)

THEME 13 EXPLAINING HOW MATERIALS BEHAVE

This is the last theme, and it is a very important one because it links together the ideas in several units which concern the reasons why materials behave the way they do. To revise this theme, you will need to refer to your notes and the *In briefs* for the units CONSTRUCTION MATERIALS, MAKING USE OF OIL, BURNING AND BONDING, THE ATMOSPHERE and SEEING INSIDE THE BODY.

What do you know about EXPLAINING HOW MATERIALS BEHAVE?

Try the following questions. Check your answers by referring to the *In briefs* for the unit.

1 Fill in the gaps in the table:

Type of structure	Building block	Example
giant lattice	atoms ions ?	diamond ? copper
molecular lattice	?	?

2 Which of the following contain molecules? Which contain ions?

(a) glass
(b) metals
(c) bricks
(d) wood
(e) oil
(f) detergents
(g) salts

3 People can use ordinary salt (sodium chloride) and Lo-salt (potassium chloride) to season food. Why are these substances quite similar?

4 Atoms contain even smaller particles called *sub-atomic particles*. What are the names of the three most common of these particles?

5 The tyres on a bicycle often feel harder in the middle of the afternoon than they felt in the morning, when it was colder. Use the kinetic model of gases to explain this.

Getting to grips with EXPLAINING HOW MATERIALS BEHAVE

Activity 1
Start by identifying the sections in the *In briefs* which relate to this theme. The key to identifying these sections is that they are the ones which use ideas about particles to help explain how materials behave.

Activity 2
Now decide on your headings for each of these sections. Again, because there are lots of ideas in this theme, and several different ways of grouping them, you might want to do this on pieces of card. This is also a theme where it would be particularly useful to compare your key words and phrases with those of a friend.

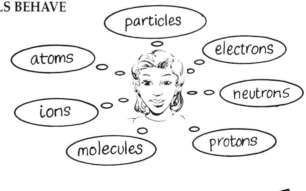

SUBSTANCES MADE FROM ATOMS
SUBSTANCES MADE FROM MOLECULES
SUBSTANCES MADE FROM IONS
THE PERIODIC TABLE
SUB-ATOMIC PARTICLES
IONIZING RADIATIONS
THE KINETIC MODEL

Activity 3

As one last task, you might find it useful to number all your summaries and produce a list of each of the summary headings. This could be in alphabetical order, or in the order you have made the summaries. This will act as an index and help you find a particular summary as quickly as possible.

And that's it!

You have now made summaries about all the ideas in the twenty-two units you have studied. In doing this you have thought hard about all these ideas. This means you should have understood them and will remember a lot about them. You have also learned how to tackle different sorts of exam questions.

Refer to your summaries whenever you have a few minutes to spare. You should find them particularly useful the night before the exam, when your last-minute revision can be a quick scan through your summaries. You are now well on the way to doing yourself justice in your exams.
GOOD LUCK!

Questions on EXPLAINING HOW MATERIALS BEHAVE

Here is another example of a free response question for you to try.

> Look carefully at the question for clues – you will need to show, in words and diagrams, that you understand that density is related to mass in a particular volume.

(a) Polythene is made by polymerizing ethene molecules. One process for making polythene uses high pressure. This makes polythene in which the polymer chains have branches. This is called low density polythene.

Another process uses a catalyst and low pressure. This makes polythene in which the polymer chains are unbranched, and it is a high density form of polythene.

Explain, using diagrams, how the structure of these two forms of polythene cause them to have different densities. (4)

(b) A group of students was asked by their teacher to investigate the physical properties of two materials and put forward theories about the types of particles and structures in the materials which would result in these properties.

The table below shows the results of their investigations.

Material	State	Appearance	Smell	When placed in electrical circuit
A	solid	hard, shiny	none	conductor
B	liquid	runny	strong smell	non-conductor

(i) For each of the substances, put forward your theory about what types of particles are present and how they are linked together. (4)

(ii) For each of the substances, explain how the properties of the substance depend on the type of particles present and how they are linked together. (4)

> Put forward – just say what your ideas are here.

> Explain – now explain your ideas. For full marks you should be explaining each of your ideas in b (i) in terms of particles.

In brief

Balancing Acts

1 The environment contains many different types of plants and animals. Human activity can change the environment. This affects the other living things on the Earth – what kinds there are, where they live and how many there are.

2 Human activity can kill organisms, or even whole species. We *can* manage the environment to encourage and protect living things, instead of killing them.

Food and other goods have to be transported from place to place.

Houses are needed for people to live in.

Fertilizers improve crop yields so provide more food.

3 We need to collect information about the environment so that we can judge whether it is changing. Sensors connected to computers are an important tool in this task.

4 The environment is very complex. Improving one part of the environment can have a harmful effect on another part. Modern farming has become very efficient thanks to agrochemicals, such as fertilizers and pesticides, but these need to be handled carefully. For example, fertilizers may improve crop growth, but they can also pollute rivers and lakes if not used properly.

5 Changes in the environment can affect how quickly populations of organisms grow. The total population that an environment can support is limited by factors such as space, food and light.

6 There is a limited amount of material available on the Earth. Important materials are used again and again, as part of a **cycle**.

   Energy cannot be recycled. It is transferred between organisms as part of a **chain**.

7 Organisms can be classified as **producers** or **consumers**. Green plants **produce** food by photosynthesis. Animals then **consume** this food, either by eating plants or by eating animals that have eaten plants.

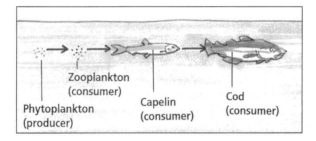

9 All the animals in a food web get their energy from the Sun, via a food chain. Not all the energy which enters a food chain is transferred to the final predator. Energy is used in the chain by the other organisms, for example, for moving, excreting or reproducing. You can draw pyramids **of numbers** and **pyramids of biomass** to represent feeding relationships which take account of this energy use.

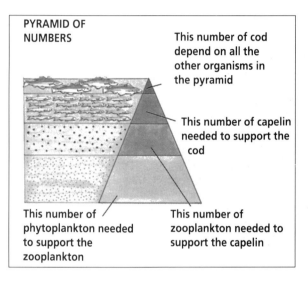

8 Animals that eat other animals are called **predators**. The animals that they eat are called **prey.** You can draw a **food chain** to show the relationship between predators and prey. **Food webs** are more complicated and show the relationships between different food chains.

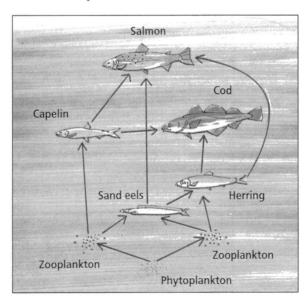

10 The oceans are a rich resource of food, but we need to manage fishing carefully. If we upset the balance, stocks of fish may be wiped out.

11 The planet Earth is a finite resource. It is made up of different environments that all affect one another. We need to manage all these environments carefully to avoid destroying them. We may have to pass laws to protect the environment.

In brief

Burning and bonding

1 Burning fuels

(a) Petrol and diesel fuel are mixtures of **hydrocarbons.** Natural gas (methane, CH_4) and bottled gas (propane, C_3H_8 or butane, C_4H_{10}) are also hydrocarbons.

(b) These fuels belong to a series of hydrocarbons called **alkanes.**

(c) When hydrocarbon fuels burn in a good supply of air, carbon dioxide and water are formed.

(d) The **flashpoint** of a fuel is the lowest temperature at which its vapour can be ignited.

(e) A certain amount of energy has to be transferred to the molecules before they will react. This is called the **activation energy.**

(f) During burning bonds are **broken** in the **reactants** (methane and oxygen) and new bonds are **made** to form the **products (carbon dioxide and water).**

(g) Breaking bonds in the reactants uses energy. **Making new bonds in the products releases energy. More** energy is released making the products than is used to break the bonds in the reactants. This means that overall the reaction is **exothermic.**

2 The effects of burning fuels on the atmosphere

The diagram shows the major effects on the atmosphere of burning fuels.

Carbon dioxide and unburnt hydrocarbons are greenhouse gases.

Sulphur dioxide and oxides of nitrogen are acidic and cause acid rain.

Carbon monoxide, oxides of nitrogen and lead compounds are particularly poisonous.

Using hydrocarbon fuels for heating or to drive an engine produces a range of atmospheric pollutants. Some of these are formed because the fuels contain small amounts of other substances or additives.

Catalytic convertors in exhaust systems convert carbon monoxide and hydrocarbons to carbon dioxide and water, and oxides of nitrogen to nitrogen.

3 Theories of bonding and structure

Most minerals are compounds of metallic and non-metallic elements. Electrons are transferred from the metal atoms to the non-metal atoms forming positively and negatively charged **ions**. The attraction between these ions is called **ionic bonding**. It holds the particles in the crystals together.

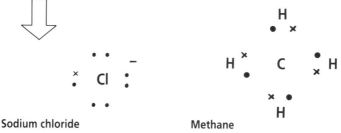

Sodium chloride

Some compounds such as hydrocarbons are made from two or more non-metallic elements. Each bond between two atoms consists of a pair of shared electrons. The atoms in the molecules are held together by **covalent bonding.**

Methane

You can use theories of bonding and lattice structure to explain many of the physical properties of solids.

A molecular lattice consists of separate molecules held together by weak forces. The forces within the molecules are stronger covalent bonds.

Ice consists of a lattice of water molecules held together by weak forces. The covalent bonding within the water molecules is strong.

A giant lattice consists of either atoms or ions held together in a giant structure.

Diamond consists of a giant lattice of atoms of carbon, each one covalently bonded to four others.

A crystal of salt (sodium chloride) consists of a giant lattice formed by positively charged sodium ions attracted to negatively charged chloride ions.

A metal consists of a giant lattice of positively charged ions held together by electrons which are spread throughout the metal.

4 How the periodic table links properties of elements and their electronic structures

Li 2,1		F 2,7	Ne 2,8
Na 2,8,1		Cl 2,8,7	Ar 2,8,8
K 2,8,8,1		Br 2,8,18,7	Kr 2,8,18,8

The vertical columns of elements in the periodic table are called groups. Elements within the same group have similar properties.

There is often a gradual variation in properties as you go up or down a group.

Elements in the same group have similar properties because they have the same number of outer electrons.

You can link the trends in properties within a group to the changes in the size of the atoms as you go up or down the group.

In brief

Communicating Information

1 We have five senses to receive messages. Other animals use their senses in different ways.

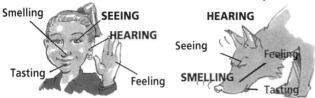

2 We use language for communicating. When we write a message, we use letters – a sort of code – to represent the spoken sounds.

3 If we want to communicate quickly with people who are far away, we convert speech into coded electrical signals. These can be transmitted over long distances along telephone wires or carried by radio waves.

4 You make spoken sounds in the voice box in your throat. You receive sounds with your ears. Hearing defects happen when people damage some part of their ears.

5 You use your eyes to receive information in the form of writing or pictures. Your eye acts as a lens and produces a focused image of the thing you are looking at on your retina.

6 We can use additional lenses – glasses or contact lenses – to correct sight defects. Other optical instruments, like microscopes, telescopes, cameras and projectors, also use lenses to produce images.

7 Lenses produce focused images of objects. They do this by bending the light as it passes through them. The thicker the lens, the more strongly it bends the light.

8 We can use **ray diagrams** like the one here to explain how a lens forms an image, and to predict the size and position of the image.

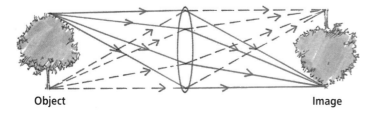

Object   Image

9 By experimenting with convex lenses, you can discover two general rules about images
   - as an object comes closer to the lens, the image moves further away from the lens, and also gets bigger
   - the stronger the lens, the closer the image is to the lens.
   These rules are true as long as the object is not too close to the lens. They are useful in understanding how cameras and projectors work.

10 If you pass white light through a prism, you can see different colours. White light is made up of these colours. The prism separates the colours in white light.

11 Light from two sources can produce an **interference pattern,** just like water waves do. This suggests that light is a sort of wave. Light consists of vibrating electric and magnetic fields – it is an **electromagnetic wave.** Electromagnetic waves can travel through empty space.

12 Different coloured lights have different **wavelengths.** Red light has a longer wavelength than blue light.

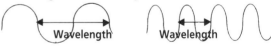

Wavelength   Wavelength

13 There are other electromagnetic waves with wavelengths shorter than blue light and longer than red light. You cannot see these waves. We use them in various ways, for example, radio communication and taking X-rays. Electromagnetic waves all travel at the same speed, the speed of light (300 000 000 m/s).

14 Light rays change direction when they pass from air into a clear medium like glass, or from glass to air. This is called **refraction.** Lenses and prisms work by refraction.

15 Light rays which strike the surface inside a clear medium may be reflected back into the medium. This is called **total internal reflection.** It is used in optical fibres. Even if the fibre is bent, light inside cannot escape and passes all the way along. Optical fibres can be used to carry coded information in light beams, instead of using electrical signals in wires.

In brief

Construction Materials

1 **Bricks:**
   - are made from fired clay
   - are strong and do not absorb much water.
   - The atoms in clay are bonded together in layers. During firing the layers become bonded together to form a three-dimensional structure.

2 **Metal ties:**
   - are made from galvanized iron.
   - Galvanizing is one way of preventing rusting.
   - Both water and oxygen are needed for rusting.

3 **Plastics:**
   - are made by polymerizing certain compounds obtained from crude oil
   - are light and weather resistant.
   - Some can be reshaped when heated — **thermoplastics**.
   - Some cannot be reshaped once formed — **thermosets**.
   - During polymerization, large numbers of small molecules (**monomers**) combine to form a very large molecule (**polymer**).

4 **Aluminium:**
   - is resistant to corrosion because of the layer of aluminium oxide on its surface.

5 **Glass:**
   - is made by heating sand, soda ash and limestone together
   - has a three-dimensional structure and is hard. But, like a liquid, it does not have a regular structure
   - can be made safer by heat-toughening or laminating with plastic.

7 **Paint:**
   - is a mixture of a polymer, a solvent and a pigment
   - is used to protect wood from rotting and metal from corroding
   - sets when the solvent evaporates and the polymer chains react with oxygen from the atmosphere to bond together.

6 **Wood:**
   - is hard, but can be split along the grain and will absorb water and rot
   - has long thin molecules lined up along the grain with weak bonds between them.

In brief

Controlling Change

1 Something that affects an organism is called a **stimulus.** The change the stimulus produces in the organism is called a **response.**

2 It is natural for changes to happen inside and around an organism. Usually changes are **controlled.** Disorder results from uncontrolled change.

3 You have two systems for controlling change: a **nervous system** and a **hormone system.**

4 The nervous system sends electrical pulses around the body. It works like a telephone – a message is sent direct to one part of the body which responds. The hormone system sends chemicals around the body. It works like a tannoy – a message is broadcast to many parts of the body, not all of which respond.

5 Change in plants is controlled by hormones. Plants respond to a stimulus by growing towards or away from the stimulus. These responses are known as **tropisms.**

6 **Phototropism** is a plant growth response to light. Positive phototropism means growing towards the light, as shoots do. Negative phototropism means growing away from the light, as roots do.

7 **Enzymes** control the chemical reactions going on in the cells of all living organisms. Respiration is one of many cellular reactions controlled by enzymes.

8 Some cellular reactions are linked to others, so that the end product of one becomes the starting point of another. Each reaction in such a chain reaction is controlled by an enzyme. The whole chain is controlled by **feedback** – some products of the reactions prevent or allow other reactions in the chain.

9 Enzymes are important in the control of digestion. Many enzymes work together to change the food we eat into the soluble substances we need. The activity of these digestive enzymes is controlled by levels of acidity or alkalinity in our alimentary canal.

Enzyme	pH2 (acid)	pH8 (alkaline)
Amylase	Inactive	Changes starch to maltose
Pepsin	Changes protein to polypeptides	Inactive

10 Enzymes work best in stable conditions of pH and temperature. Many organisms can control the conditions inside their bodies. Keeping internal body conditions stable is called **homeostasis.** The better the homeostatic system the better the enzyme activity and the more healthy the organism.

11 Mammals and birds have good homeostatic systems. They are able to cope well with changes in their external environment and so are found in a wide range of habitats.

12 The movement of water through a plant is called the **transpiration stream.** A plant gains water through hairs on its roots and loses it through holes called **stomata** in its leaves. Transpiration happens when water evaporates through the stomata. This evaporation cools the leaf.

   Plants can open and close their stomata which affects their rate of transpiration.

A large oak tree can lose more than 600 dm³ of water per day through transpiration.

In brief

Electricity in the Home

1 There are two types of electric charge which we call **positive** and **negative**. Two objects with the same charge repel each other (they move apart). Two objects with opposite charges attract each other (they move together).

2 You can **charge** insulators by rubbing them with a cloth. Electrons are rubbed off one material (leaving it with a positive charge) and stick on to the other (giving it a negative charge).

3 You can also charge conductors by rubbing them, if they are insulated from earth. If they are connected to earth the charge immediately escapes. Sometimes it escapes as a spark – a sudden discharge of static electricity through the air.

4 An electric **current** is a flow of charge. In a wire, the flowing charges are electrons.

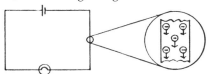

The size of the current depends on how much charge flows past a point each second:

$$\text{current} = \frac{\text{charge}}{\text{time}}$$

We measure current in amperes (or **amps**) and charge in **coulombs**. A current of 1 amp means that 1 coulomb of charge is passing each second.

5 Electric current can only flow round a closed conducting loop called a **circuit**. A source of electrical energy (a cell, battery or mains power supply) makes the current flow. The current is the same at all points round the loop. The circuit transfers energy, for example from a battery to a bulb. But the current which carries the energy is the same before it gets to the bulb (at A) as it is after (at B).

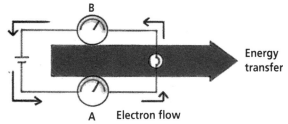

Electron flow

6 You can vary the size of the current in a simple circuit by changing the battery **voltage** and the **resistance** in the circuit. The voltage (*V*) is the electrical push which drives the current round the circuit. It is measured in **volts**. The bigger the battery voltage, the bigger the current.

A light bulb, resistor or other device in the circuit has **resistance**. With any particular battery, the bigger the resistance, the smaller the current. Resistance (*R*) is measured in **ohms** (Ω).

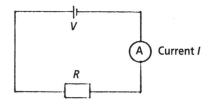

The circuit equation summarizes all this. It can be written as

$$V = IR$$
$$\text{or } R = \frac{V}{I}$$
$$\text{or } I = \frac{V}{R}$$

where V = voltage
R = resistance
and I = current

7 You can use the circuit equation to find the resistance of a component. You measure the current and voltage and from this you can calculate its resistance.

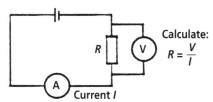

Calculate:
$$R = \frac{V}{I}$$

8 You often want to use more than one device in the same circuit, for example you might want to use two light bulbs or two resistors. You can connect them in different ways.

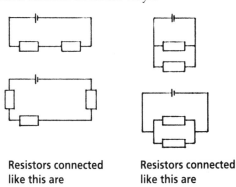

Resistors connected like this are **in series**.

Resistors connected like this are **in parallel**.

▶

9 The devices in most of the circuits you meet in everyday life are connected in parallel. This is because, in parallel circuits

- if one device fails, the others continue to work

- you can switch each device on and off independently

- each device operates off exactly the same voltage, the voltage of the supply.

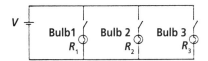

In this simple parallel circuit, you can turn each bulb on and off independently.

10 You can work out the total current drawn from the supply by several devices in parallel. To do this you need to look at each device separately.

   Using the circuit equation, the current through bulb 1 above is $\frac{V}{R_1}$. You can work out the currents through bulbs 2 and 3 in the same way.

   You can find the total current from the supply by adding the currents drawn by each bulb.

11 If you connect components such as resistors in series, you can find the total resistance by adding the resistances. And the voltages across the resistors add up to the voltage of the battery.

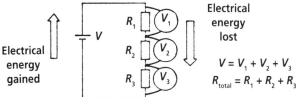

$$V = V_1 + V_2 + V_3$$
$$R_{total} = R_1 + R_2 + R_3$$

The voltage gives you an idea how much energy the electrons gain and lose as they flow round the circuit. Electrons gain electrical energy from the battery and lose it again in the resistors.

   The voltage is the number of joules of energy gained or lost by each coulomb of charge.

$$\text{voltage} = \frac{\text{energy}}{\text{charge}}$$

$$V = \frac{E}{Q}$$

12 The energy transferred every second by an electrical circuit is called **power**. It depends on the current and voltage in the circuit.

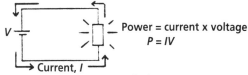

Power is measured in **watts** (W).

13 If you know the power rating of a mains device, such as an electric fire, you can use the power equation above to calculate the current it uses. The mains voltage is 240 V.

$$\text{current (in A)} = \frac{\text{power rating (in W)}}{240}$$

The cable and the fuse used for the fire must be capable of carrying this current. They must have a higher current rating than this.

14 Chemical reactions often give out energy. In **electric cells** chemical energy from reactions is transformed into electrical energy.

15 Electric cells have two electrodes made of metal or carbon. In the chemical reaction, one electrode (the negative electrode) loses electrons. These electrons flow along wires through the external circuit to the positive electrode. The chemical reaction at the positive electrode 'mops up' these electrons. Between the electrodes is the **electrolyte**, which might be an acid, an alkali or a salt solution. In **dry cells** the electrolyte is a paste.

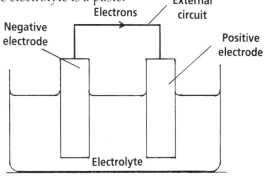

16 Two or more cells joined together make a battery. **Primary** cells cannot be recharged after they run down. You can recharge a **secondary** cell, such as a lead–acid cell, by passing electric current through it. This converts electrical energy back into chemical energy.

In brief

Energy Matters

1 Our comfortable lifestyle depends on using a lot of energy, at home and at work.

2 Fuels are concentrated stores of energy.

Primary energy sources

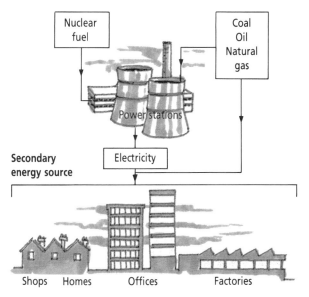

3 Coal, oil and natural gas are fossil fuels. They are the fossilized remains of forests and of small sea creatures which lived millions of years ago. Once we have used them up, they are gone for ever.

4 Different tasks take different amounts of fuel. If we want to measure the amount of energy used, we need a common unit for energy. This is the joule (J).

5 Some electrical appliances use energy more quickly than others. Those which involve heating use a lot of energy.

6 The electricity meter in your home measures how much electrical energy you are using. It 'adds up' the energy used by all your appliances.

7 Every domestic electrical appliance has a power rating in watts (W) written on it. The bigger the power rating, the faster it uses energy. An appliance with a large rating will cost more to run than one with a smaller rating if they are kept switched on for the same time.

8 Appliances do not really 'use' energy – they change it from one form to another, or transfer it from one place to another. This can be illustrated by an energy arrow diagram.

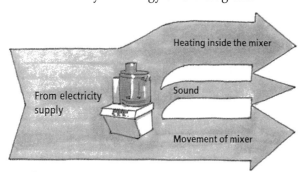

9 There are two important energy *laws*:
   ● There is always the same total amount of energy after an event as there was at the beginning (**conservation of energy**).
   ● Energy always spreads out from concentrated sources (fuels) and ends up in many places. Some always ends up causing unwanted heating (**spreading of energy**).

10 Hot objects cool down and their surroundings get slightly warmer. The energy in the hot object has spread further.

ENERGY SPREADS BY . . .

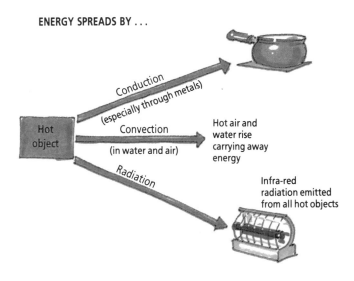

11 We can reduce our fuel bills by insulating our homes better. Insulation makes it harder for energy to spread.

12 Energy is needed to *make* things. Everything has an 'energy cost' as well as a 'raw materials cost'.

In brief

Energy Today and Tomorrow

1 Our way of life depends on using **fuels**. **Fossil fuels** (coal, oil and natural gas) and **nuclear fuels** (uranium and plutonium) provide a concentrated source of energy.

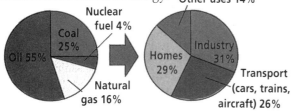

Sources of our energy

How we use energy

2 The Earth's reserves of fossil fuels and nuclear fuels are limited. They will not last forever. So it is important to conserve fossil fuels, by

- using fuels efficiently and avoiding waste
- developing alternative sources of energy, especially renewable sources.

3 When a fuel is used, its energy spreads out and becomes **dispersed**.

4 An electric current is generated when a wire moves through a magnetic field. This is called **electromagnetic induction**. The induced current is a **direct current**. If the wire moves in the opposite direction, the direction of the current changes.

5 A coil rotating in a magnetic field generates an **alternating** electric current.

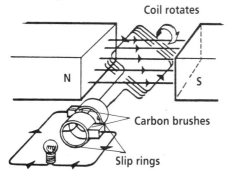

The current can be made bigger by
- using a stronger magnetic field
- rotating the coil faster
- using a coil with more turns
- winding the coil on a soft-iron core.

6 Most power stations use the energy in a fossil fuel or a nuclear fuel to boil water and make steam. This turns turbines which rotate a generator coil to produce electricity. About one-third of the energy stored in the fuel is turned into useful electricity; the rest is wasted in the hot flue gases and hot cooling water.

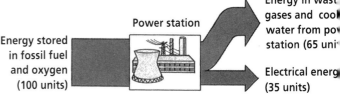

7 When a fossil fuel burns, it combines with oxygen in the air. This is a chemical process called **combustion**. Energy is released. Carbon dioxide and water are formed, along with other unwanted substances.

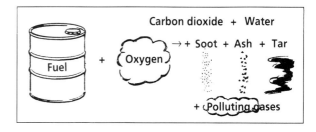

8 **Pollution** is the contamination of the environment by substances made by humans. Combustion of fuels often causes air pollution.

Pollution must be cut down, and this costs money. We have to balance this cost against the benefits of controlling pollution.

9 Many fuels, particularly coal and coke, contain sulphur. When they burn, the acidic gas sulphur dioxide is formed. This can help form **acid rain**.

Acid rain is harmful to living things such as trees and life in rivers and lakes. Acid rain and sulphur dioxide in the air cause damage to buildings. They make metals corrode faster.

10 Most of the sulphur dioxide can be removed from power station flue gases by a **flue-gas desulphurization** (**FGD**) plant, but this is expensive to install and run. It requires large amounts of limestone and produces large quantities of gypsum.

11 A **nuclear fuel** contains atoms with large, unstable nuclei. If a neutron hits one of these nuclei, the nucleus splits in half, releasing energy. This splitting is called **fission**. Several neutrons are also emitted, which can cause other fissions. This may lead to a **chain reaction** in which large amounts of energy are released.

12 There are both benefits and drawbacks to nuclear power.

Benefits

(a) Nuclear fuels will last for another 2000 years.

(b) Nuclear fuels produce no smoke, no soot and no acid rain.

(c) Unlike fossil-fuelled power stations, nuclear power stations do not emit carbon dioxide. The amount of this gas in the atmosphere is increasing, and this may be causing global warming.

(d) Nuclear fuel saves valuable fossil fuels for other uses.

Drawbacks

(a) Nuclear power stations involve risks from accidents with radioactive substances.

(b) There are dangers from leaks of radioactive substances from nuclear power stations.

(c) Some of the waste from nuclear reactors will be radioactive for hundreds of years and is extremely difficult to dispose of safely.

13 **Renewable energy sources** will last as long as the Earth itself. Their supply is unlimited.

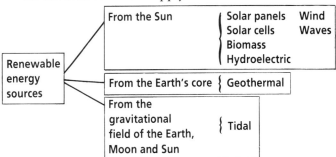

14 Renewable energy sources are difficult to **harness**. Their energy is very 'spread out' or dispersed. It is difficult to concentrate this energy to produce high temperatures or to generate electricity.

15 **Nuclear fusion** may be the energy source of the future. When two small nuclei of light elements are joined together (fused), energy is released. Fusion reactions inside the Sun provide the Sun's energy but they are proving very difficult to harness on Earth.

16 A **transformer** is used to change the voltage of an electricity supply. It consists of two separate coils wound on the same iron core. When the current changes in one coil (the **primary**) it induces a current in the other coil (the **secondary**). An alternating current in the primary induces another alternating current in the secondary.

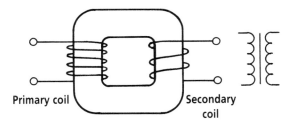

Primary coil   Secondary coil

The secondary voltage can be calculated from the equation:

$$\frac{\text{secondary voltage}}{\text{primary voltage}} = \frac{\text{number of turns in secondary coil}}{\text{number of turns in primary coil}}$$

or voltage ratio = turns ratio

17 If the secondary has more turns than the primary, the transformer is a **step-up** transformer. If the secondary has fewer turns, it is a **step-down** transformer.

18 In a step-up transformer, the output (secondary) voltage is larger than the input. But the output *current* is lower than the input current – by the same ratio. This means that the input and output power is the same (as $P=IV$) – in a 'perfect' transformer. In a real transformer, there is some loss of power.

19 The National Grid is a network of cables and wires which links all the power stations in Britain to the users of electricity. Transformers step up the voltage from the power stations to over 400 000 V for long-distance transmission. This means that the current in the cables is low and less energy is lost in transmission.

In brief

Evolution

1 Individuals of a species show variation in their **characteristics**, such as hair colour or eye colour.

2 The characteristics of an individual are determined by the interaction of **inherited factors** and **environmental factors**.

Soil A   Soil B

Plants with the same genetic factors grow differently in different soils.

Soil A   Soil A

Plants with different genetic factors grow differently in the same soil.

3 Individuals with characteristics best suited to their environments are most likely to survive and reproduce. This is called **natural selection**. For example, a moth the same colour as its surroundings is less likely to be eaten than one with a contrasting colour. The offspring of the camouflaged moth are likely to be camouflaged as well. Over a period of time, the proportion of camouflaged moths increases. The characteristics of the population change. This change or **evolution** is due to the **survival of the fittest**.

4 The theory of evolution by natural selection explains how organisms can change over very long periods of time, and how present-day organisms could have developed from earlier forms.

5 Selective breeding is used to produce plants and animals with desirable characteristics. This is called **artificial selection**.

6 Fossils give evidence about the types of organisms that existed in the past.

7 There are conflicting theories on the origin of present-day lifeforms.

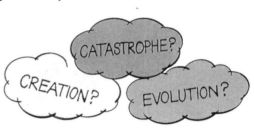

8 Cells contain thread-like structures called **chromosomes**. These carry coded information needed for the control and development of the organism.

9 All the cells of an individual come from one original cell. This is created when the male and female parent cells join. Each parent cell contributes half the final number of chromosomes to the new cell.

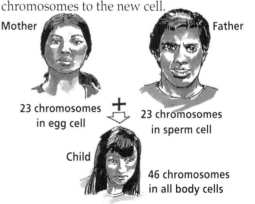

Mother   Father

23 chromosomes
in egg cell

23 chromosomes
in sperm cell

Child

46 chromosomes
in all body cells

10 Chromosomes carry **genes**. A gene is responsible for the inheritance of a particular characteristic.

Different forms of a gene are called **alleles**. Different alleles cause different characteristics. For example, some people can roll their tongue and others cannot. These are two alternative characteristics. People who can roll their tongue have the tongue-rolling allele, those who cannot have the non-tongue-rolling allele. Both are alleles of the tongue-rolling gene.

11 Chromosomes occur in pairs. The chromosomes in these **homologous** pairs look the same as each other and carry the same genes. However, each chromosome may have a different allele for a particular gene.

Two homologous pairs of chromosomes

12 Within a pair of alleles, one is usually **dominant**. This means that if both alleles are present in an individual's chromosomes, the individual shows only the dominant characteristic. The allele for the hidden characteristic is called the **recessive** allele. For example, in peas, the tall allele is dominant over the short allele, which is recessive.

   When neither allele is dominant, there is a mixing of characteristics. This is called **co-dominance**. In purebred Longhorn cattle, crossbreeding a white cow with a red bull results in roan calves with a mixture of red and white hairs.

13 It is possible to identify patterns in the way certain characteristics are inherited. From these patterns we can predict the chances of a characteristic being inherited in the offspring of certain parents.

14 Whether you are male or female is determined by the genes on your **sex chromosomes**. Females have two X chromosomes. Males have one X and one Y chromosome.

Sex chromosomes

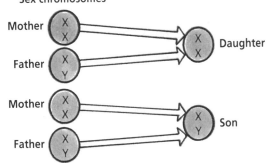

15 Characteristics controlled by genes, such as fur colour, are **inherited. Acquired** characteristics, such as large muscles resulting from bodybuilding, are not inherited.

16 Chromosomes contain the chemical **DNA**. A molecule of DNA consists of two long chains. These chains are twisted together to form a shape called a **double helix**.

Part of a DNA molecule

17 Genes are particular segments of the DNA chain. The positions of particular genes can be mapped along the length of a chromosome.

18 The DNA chain can be changed in such a way that genes on it change. These changes are called **mutations**. Mutations can result in changed characteristics. For example, a mutation which produces an extra chromosome in humans causes Down's syndrome.

19 Mutations are natural events. Not all mutations are harmful. Some chemicals and other environmental factors such as radiation increase the rate of mutation.

20 There are techniques which can deliberately alter an organism's DNA and so change its characteristics. These are called **genetic engineering**.

21 Large numbers of genetically identical individuals can be raised from a single parent. This is called **gene cloning**.

22 Genetic engineering and gene cloning have resulted in benefits for humans. However, great care has to be taken to ensure that genetically engineered organisms are not accidentally released into the natural environment.

Diabetic people have to inject themselves with the hormone insulin. Before genetically engineered human insulin was available, insulin from pigs and cattle was used.

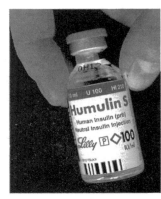

In brief

Food for Thought

1 People have increased world food production by:

Using chemicals, such as pesticides and fertilizers

Making better use of lands, such as deserts or marshes

Making more use of machines for farming and transport

Using better adapted varieties of plants

WINTER WHEAT

Making available supplies of water

2 Growing and harvesting crops removes nutrients from the soil which are essential for healthy growth.

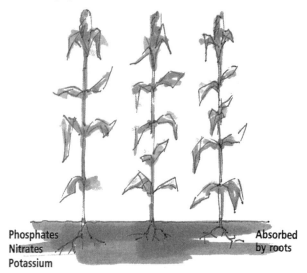

Phosphates
Nitrates
Potassium

Absorbed by roots

Nutrient-deficient soil can be improved by adding chemical fertilizers. Most chemical fertilizers contain compounds of nitrogen, potassium and phosphorus.

3 Compounds of nitrogen are particularly good for encouraging plant growth. Although there is plenty of nitrogen gas in the air, most crop plants cannot use it directly. But the nitrogen can be combined with hydrogen to form ammonia. The industrial process developed by Fritz Haber is used to do this.

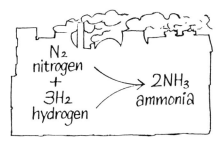

N_2 nitrogen + $3H_2$ hydrogen → $2NH_3$ ammonia

4 Ammonia can be used as a fertilizer, but it is a **base** — it dissolves in water to form an alkaline solution. This means it would change the pH of the soil. It is also smelly and difficult to handle.

This problem is overcome by reacting ammonia with acids to form **salts.** These are solids and almost neutral. This reaction is called a **neutralisation.**

5 Fertilizer manufacturers need to know how much of each chemical to react together to produce the required quantity of the fertilizer. This can be calculated from the balanced equation for the reaction and the relative molecular masses of the substances involved.

$$2NH_3 + H_2SO_4 \rightarrow (NH_4)_2SO_4$$
RMM34   98   132

34 tonnes of ammonia and 98 tonnes of sulphuric acid give 132 tonnes of ammonium sulphate.

Relative molecular masses are calculated from the relative atomic masses of the elements in the molecule.

H_2SO_4 Relative atomic mass

$H_2 = 1 \times 2 = 2$
$S = 32 \times 1 = 32$
$O_4 = 16 \times 4 = 64$
Relative molecular mass = $\underline{98}$

6 Because salts used as fertilizers are soluble, they may be washed or **leached** out of soil and reach streams and rivers, causing pollution. For example, an increase in the nitrate concentration of a river can result in unwanted growths of algae and a lower quality of drinking water.

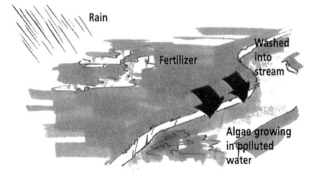

Rain

Fertilizer

Washed into stream

Algae growing in polluted water

7 To keep our stored food fit to eat we need to prevent the growth of microbes and keep out animal pests. Many animal pests are controlled by chemical pesticides, but non-chemical means can also be used. For example, instead of spraying aphids (greenfly), ladybirds can be released on the crop to eat the aphids. Some foods have preservatives added to them to make them keep longer.

Locusts can devastate crops. Chemical sprays are used to kill them.

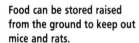

Food can be stored raised from the ground to keep out mice and rats.

8 Food can be processed to make it more useful or attractive. This can be done by
   ● chemical reactions, such as adding hydrogen to vegetable oil to make margarine
   ● biological reactions, such as using the enzyme pectinase to extract juice from apples
   ● food additives which improve the flavour or appearance of the food.

Recently people have found that some food additives can be harmful. However, foods go 'off' much more quickly without preservatives.

In brief

Keeping healthy

1 Many complex chemical reactions go on inside your body. If any of these reactions goes wrong, you become ill. This may lead to unusual chemicals in your urine, so examining urine can help in the diagnosis of illness. Reactions can go wrong for various reasons:

2 **Microbes** are tiny organisms that you need a microscope to see. **Bacteria** and **viruses** are microbes. There are many different types of microbes around us all the time, and your body makes a good environment for them to grow in!

   Pathogenic microbes cause illness when they grow in the body. They can be passed on to other people and so infect them as well.

How microbes enter the body

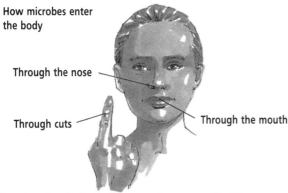

Through the nose

Through cuts

Through the mouth

3 A microbial disease can be controlled by destroying the microbe or by eliminating the conditions it needs to grow. Microbes can be killed by:
   ● high temperatures
   ● acid or alkaline conditions
   ● removal of nutrients
   ● chemicals that poison them. These are called **germicides.**

4 **Disinfectants** contain concentrated germicides. They are only used on non-living material. **Antiseptics** are weak germicides that can be used on your skin. **Antibiotics** are germicides that are safe enough to be eaten or injected into the body.

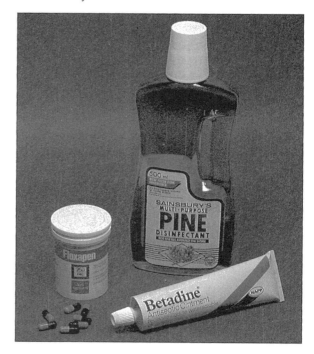

5 Your skin is a barrier to microbe invasion. When your skin gets cut, a blood clot forms to seal the wound and stop microbes entering.

6 When microbes do get into your body they are dealt with by the immune system. White blood cells produce antibodies which destroy the microbes. The next time you are infected with the same microbe, you already have antibodies to destroy it – you are **immune** to it.

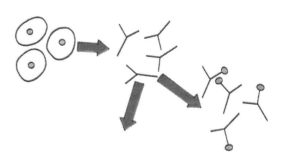

You become immune to a particular disease if you are vaccinated with a weakened form of the microbe. This stimulates the body to produce antibodies without making you ill. Immunization programmes have eliminated some major diseases.

7 You can also become immune by injection of antibodies.

8 AIDS (acquired immune deficiency syndrome) results from infection by the human immuno-deficiency virus (HIV). HIV attacks white blood cells and blocks the immune system by preventing antibody production. AIDS sufferers are unable to offer any natural resistance to microbial infection.

9 When body organs become diseased or do not function properly, they can often be replaced by healthy transplants. Kidneys, livers, lungs, hearts and corneas (from the eye) have all been transplanted. Although transplanted organs are carefully selected, they are often rejected by the body because of the immune response.

10 Your kidneys are high-pressure filter systems which control the amounts of water and dissolved substances in the blood. Although you are born with two kidneys, you can have a full and active life with only one. Kidney failure can be treated by regular dialysis (filtering) on a kidney machine, or by transplanting a healthy kidney from a donor.

11 Coronary heart disease happens when the blood vessels which supply the heart with oxygen get blocked. It can be treated by bypass surgery. An artery from the leg is transplanted to provide an alternative route to the heart.

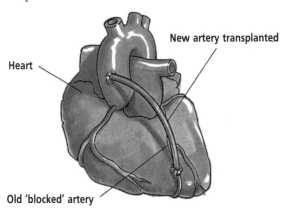

Heart

New artery transplanted

Old 'blocked' artery

Valves in the heart keep the blood flowing in the right direction. Damaged heart valves may be replaced by artificial ones. Irregular heartbeats can be corrected with an electronic pacemaker.

12 **Enzymes** (biological catalysts) control the complex reactions that go on inside your body. To stay healthy, your body must keep the right conditions for enzymes to work in.

13 The chemical reactions which enzymes catalyse take place at an active site on the enzyme molecule. A substance which has a molecular structure that fits the active site can alter the action of an enzyme. Some drugs work by blocking the active sites of enzymes.

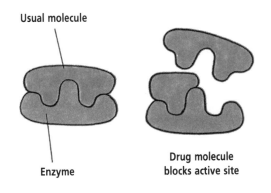

Usual molecule

Enzyme

Drug molecule blocks active site

A drug that blocks the active site of an enzyme in a disease-carrying microbe will kill the microbe.

In brief

Making Use of Oil

1 Oil has been formed from the remains of tiny water creatures that died millions of years ago. It is found trapped in porous rocks which are covered with a layer of non-porous rock.

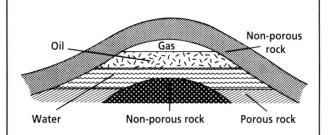

2 Oil is a complex mixture of **hydrocarbons** – compounds made up of carbon and hydrogen. It is **refined** (separated) by first distilling it and then collecting fractions which boil at different temperature ranges.

During distillation the forces between molecules are overcome.

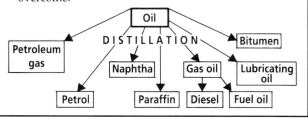

3 Certain fractions of oil are used as fuels. Petrol and diesel are used in engines and fuel oil is used as a source of heat energy for central heating systems and industry. When a hydrocarbon like petrol or fuel oil burns in a good supply of air, carbon dioxide and water are formed. For example:

$$CH_4 + 2O_2 \quad CO_2 + 2H_2O$$
the simplest
hydrocarbon

4 Other fractions of oil are **cracked**. This means large molecules are broken down into smaller molecules. The molecules are cracked by heating them in the presence of catalysts. ▶

▶ During cracking bonds between atoms within molecules are broken.

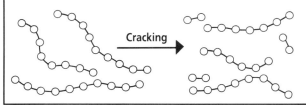

5 **Alkenes** are important products of cracking. Ethene, C_2H_4, is the simplest alkene. It can be polymerized to form polythene or poly(ethene), which is an important plastic. The molecules of ethene (**monomers**) are added together to form larger molecules (**polymers**), so polythene is called an **addition polymer**.

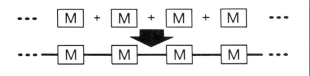

6 During some other polymerizations the polymer is not the only product. As the monomers combine, another substance with small molecules such as water or hydrogen chloride is formed. This is called a **condensation polymerization**.

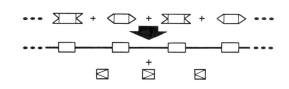

7 Non-soapy detergents are also made from chemicals which come from oil. Detergent molecules consist of a long hydrocarbon tail which is attracted to oil and grease, and a head which is attracted to water.

8 Modern society is very dependent on hydrocarbon molecules, whether as fuels or as sources of other materials such as plastics. When oil runs out we will have to use coal as a source of these.

In brief

Mining and Minerals

1 Rocks are usually mixtures of substances. We call each pure substance in a rock a **mineral**.

2 A rock which contains a useful mineral is sometimes called an **ore**.

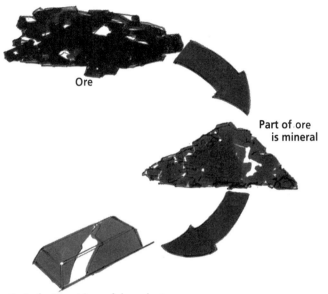

Ore

Part of ore is mineral

Part of mineral is useful product

Limestone is almost pure calcium carbonate. This means the rock which you dig out of the ground is pure mineral.
    Copper ores usually contain less than 1% of the useful mineral. The rest of the rock is useless. These are called low grade ores.

3 Mineral resources are very valuable because many products which you use every day come from minerals. But they are a finite resource which means once they have been used they cannot be replaced.

4 A pure mineral is a single chemical compound which can be represented by a formula. For example

galena is PbS

limestone is $CaCO_3$

haematite is Fe_2O_3

These formulas tell you which elements the minerals contain. They also tell you the ratio of the number of atoms of each element.

5 Often a really useful mineral is buried deep underground. Geologists use **remote sensing techniques** to find these minerals rather than go to the expense of drilling holes.

These methods include

● measuring differences in the magnetic or conductivity properties of the layers of rocks under the surface

● doing a chemical analysis of minerals dissolved in rivers and lakes

● examining photographs taken by satellites.

6 The next step after locating a mineral is to decide whether to extract it. Extraction will involve mining or quarrying. A quarry is entirely on the surface. Mines sometimes go underground.
A mining company's decision whether to extract a mineral will be influenced by:

Scientific evidence	What grade is the ore?
Technical evidence	How easy is it to extract the ore?
Market predictions	What is the likely demand for the product?
Environmental and social factors	How will it affect the environment and the local community?

7 Some of the possible benefits and drawbacks of mining are:

Benefits	Drawbacks
More jobs	Spoils appearance of countryside
Produces useful materials	Harms plant and animal life
Creates wealth for company and community	Leaves waste to be disposed of

8 Extracting a metal involves the processes shown here.

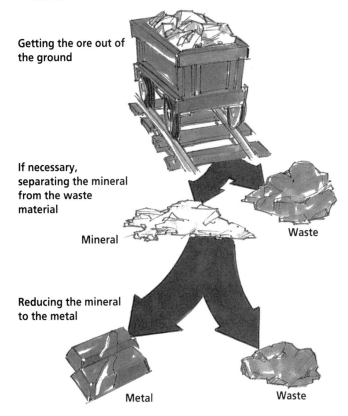

Getting the ore out of the ground

If necessary, separating the mineral from the waste material

Mineral   Waste

Reducing the mineral to the metal

Metal   Waste

Metals occur naturally in different forms.

● Iron and aluminium occur as oxides.

● Lead, zinc and copper occur as sulphides. These are first converted into the oxides, and then to the pure metal.

● Gold occurs uncombined with any other element.

● Sodium and magnesium occur as chlorides.

The method used to extract the metal depends on how reactive it is.

The less reactive metals copper, lead, iron and zinc are extracted by heating the oxide with a reducing agent such as carbon (coke).	The more reactive metals sodium, magnesium and aluminium are extracted by electrolysis.

9 During **electrolysis** an electric current is passed through an **electrolyte.**

Electrolytes are compounds which conduct electricity when they are molten or dissolved in water. They conduct because reactions take place at the **electrodes**.

For example, the electrolysis of molten sodium chloride would form sodium at one electrode

$$Na^+ + e^- \rightarrow Na$$

and chlorine at the other

$$2Cl^- - 2e^- \rightarrow Cl_2$$

10 More limestone is extracted from the ground than any other mineral. It is used direct from the quarry for building roads and for extracting iron. But for some purposes it is first heated. This converts it from calcium carbonate to calcium oxide (quicklime, or lime).

$CaCO_3$	$\rightarrow$	CaO	+	CO_2
calcium carbonate (limestone)	$\rightarrow$	calcium oxide (quicklime)	+	carbon dioxide

11 Halite, sometimes called rock salt, is sodium chloride. Chlorine and sodium hydroxide are made from it. They are used to manufacture various products including:

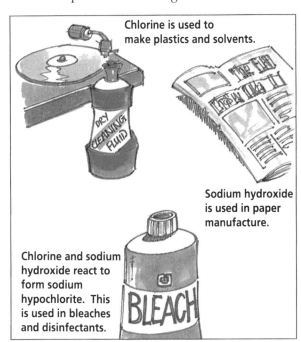

Chlorine is used to make plastics and solvents.

Sodium hydroxide is used in paper manufacture.

Chlorine and sodium hydroxide react to form sodium hypochlorite. This is used in bleaches and disinfectants.

In brief

Moving On

1 If an object is at rest (not moving), all the forces on it must be **balanced**. They add to zero.

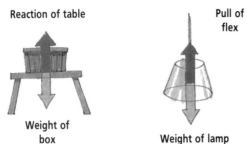

Reaction of table

Weight of box

Pull of flex

Weight of lamp

2 Forces **change** motion. Despite what you might think from everyday experience, a force is *not* needed to keep an object moving at a steady speed. This is called **Newton's first law of motion.**

Driving force

Counter force

Steady speed, so driving force = counter force

3 A force *is* needed:
   ● to start an object moving
   ● to stop an object moving
   ● to make an object move faster
   ● to make an object move slower
   ● to make an object change its direction of motion.
   But a force is *not* needed:
   ● to keep an object moving in a straight line at a steady speed.

4 Left to themselves, objects do not change their motion. This property of all objects is called **inertia.**

5 If a force acts on an object, it makes the object **accelerate.**

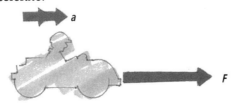

Force causes acceleration

6 Acceleration is defined by the equation:

$$\text{acceleration} \quad = \quad \frac{\textbf{change of velocity}}{\textbf{time taken}}$$

The units of acceleration are m.p.h. per second, or m/s per second.

A positive acceleration means that an object is speeding up. A negative acceleration means that it is slowing down.

7 The acceleration of an object depends on the size of the force acting on it and on its mass:

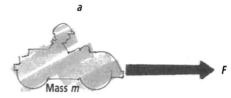

Mass m

   ● Acceleration is directly proportional to the force acting.
   ● Acceleration is inversely proportional to the mass of the object.
   This is summarised by the equation:

$$\textbf{force} = \textbf{mass} \times \textbf{acceleration}$$
$$F = ma$$

This is known as **Newton's second law of motion.**

8 The unit of force is the newton (N). One newton is the force needed to give a mass of 1 kg an acceleration of 1 m/s per second.

9 Another useful way to state Newton's second law is:

$$\textbf{force} \times \textbf{time} = \textbf{mass} \times \textbf{change in velocity}$$
$$Ft = mv_{\text{final}} - mv_{\text{initial}}$$

From this it follows that:
   ● if an object is stopped very quickly, the force involved is large
   ● if an object is stopped in a longer time, the force involved is smaller.

10 When a force acts on a surface, it exerts a **pressure.** The size of the pressure depends on the force and the area on which it acts:

$$\text{pressure} \quad = \quad \frac{\textbf{force}}{\textbf{area}}$$

11 The quantity mv is called **momentum.** When two objects collide, or spring apart (an explosion), the total momentum is the same afterwards as it was before.

In Brief

Restless Earth

1 The Earth is made up of three zones: the **core**, the **mantle** and a thin **crust** on which we live. This diagram shows the structure of the Earth and the composition of the different parts.

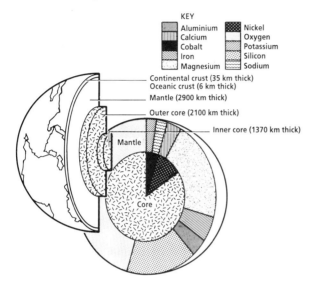

KEY
- Aluminium
- Calcium
- Cobalt
- Iron
- Magnesium
- Nickel
- Oxygen
- Potassium
- Silicon
- Sodium

Continental crust (35 km thick)
Oceanic crust (6 km thick)
Mantle (2900 km thick)
Outer core (2100 km thick)
Inner core (1370 km thick)

Mantle

Core

2 People have found out about the structure of the Earth by studying earthquakes. Earthquakes send out shock waves which can be felt at places all over the world. By recording where the waves are felt and where they are not, geologists have built up a picture of the inside of the Earth.

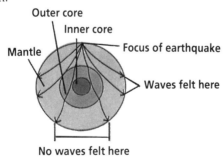

Outer core
Inner core
Mantle
Focus of earthquake
Waves felt here
No waves felt here

3 Earthquakes are caused by the Earth's crust moving. The crust is broken into different plates, which float on the liquid mantle.

There are convection currents in the mantle which move the plates. As the plates move they scrape and bump against other plates. This uneven movement causes violent movements in the surrounding rocks.

4 There are three types of rock in the crust.
Igneous rocks have been formed by volcanic action. They were molten, but have now cooled.
Sedimentary rocks have been laid down over very long periods of time. They are formed from sediments that have slowly built up at the bottom of shallow seas.
Metamorphic rocks form when heat and pressure in the Earth change igneous or sedimentary rocks into a different form.

5 Layers of rock may be bent or folded as a result of movements in the Earth.

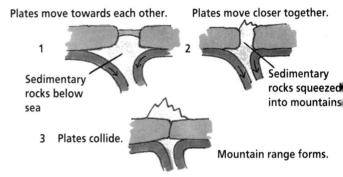

Plates move towards each other.
1
Sedimentary rocks below sea

Plates move closer together.
2
Sedimentary rocks squeezed into mountains

3   Plates collide.
Mountain range forms.

6 Sometimes folding rocks push up to produce mountains. Mountains are gradually worn away by wind and rain. The worn away rock is carried out to sea by rivers.

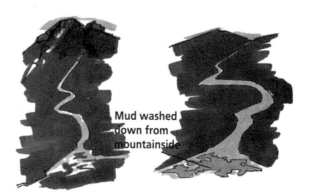

Mud washed down from mountainside

7 Occasionally, when animals or plants die, their remains are preserved as **fossils**. Fossils become part of the rock, as sediment is laid down around them.

Some organisms exist for a time and then become extinct. Knowing how long ago certain organisms lived is very useful to geologists – they can tell the relative ages of different rocks by looking at the fossils inside them.

In brief

Seeing Inside the Body

1 Doctors can use a variety of methods to find the cause of a patient's illness. **Invasive** methods put something into the body or take something out of it. **Non-invasive** methods do not.

Diagnostic methods

Non-invasive	Invasive
• Talking to the patient	• Taking a blood sample
• Tapping	• Surgery
• Feeling	• Radioactive tracers
• Listening with stethoscope	
• X-ray	
• Ultrasound scan	

2 X-rays are produced in an X-ray tube when fast-moving electrons are stopped by the metal anode. X-rays can pass easily through soft tissue (skin, muscle, fat) but not bone. So they can be used to take a 'shadow photograph' of bones.

3 **Radiation** transfers energy from one place to another. Types of radiation include:
- X-rays
- radio waves
- light
- sound waves
- infra-red

Radiation always comes from a **source** and gradually gets weaker as it spreads out.

source ———radiation———→ **detector**

You can detect some kinds of radiation with your senses, but for others you need special instruments.

4 **Radioactive substances** emit an invisible radiation which can be detected by a Geiger–Müller (GM) tube and counter.

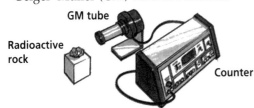

GM tube

Radioactive rock

Counter

This radiation is emitted in bursts, not smoothly and continuously. You cannot switch a radioactive source off and on like you can an X-ray tube.

5 X-rays and the radiation from radioactive substances are called **ionizing radiation.** As they pass through materials they convert atoms into ions. They do this if they pass through the cells in your body. This can damage cells. So safety precautions must be taken with ionizing radiations.

6 Materials like lead which absorb ionizing radiation can be used to screen people.

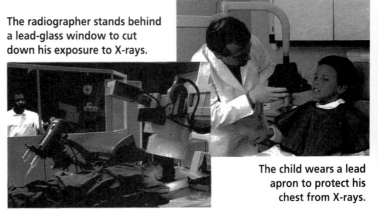

The radiographer stands behind a lead-glass window to cut down his exposure to X-rays.

The child wears a lead apron to protect his chest from X-rays.

7 There are small amounts of radioactive substances all around us – in the soil, in the air and in food. So there is always some **background radiation** from these sources.

8 The amount of ionizing radiation a person absorbs is called the **radiation dose.** It is measured in **sieverts** (Sv). The biggest contribution to people's radiation dose in the UK is from **radon**, a radioactive gas which comes out of the ground and can collect in buildings.

9 Radiation workers wear **film badges** to measure the dose they receive. This dose should be kept as low as possible and must be below the agreed safety standard.

10 There are three different types of radiation from radioactive substances.

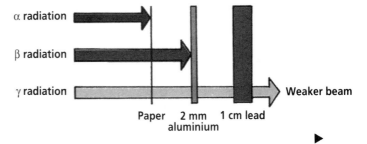

α radiation

β radiation

γ radiation   Weaker beam

Paper   2 mm   1 cm lead
        aluminium

▶

11 The **activity** of a radioactive source is the number of radioactive emissions from the source every second. It is measured in **becquerels** (Bq). One Bq means one emission per second.

12 The activity of a radioactive source decreases as time goes by, but it never drops completely to zero. It always takes the same time for the activity of a source to fall to half its original value. This time is called the **half-life** of the source. Half-lives can vary from less than one second to more than a million years.

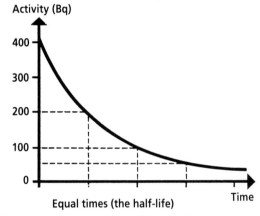

13 Radioactive sources are used in medicine in two main ways:

To treat cancer – a powerful beam of radiation from a strong radioactive source is pointed at the tumour for a short time. The radiation kills cells along its path. The treatment is repeated several times from different angles to try to kill all the cells in the tumour.

To see inside the body – a small amount of a radioactive substance is injected into the patient. Detectors can then follow the movement of the radioactive substance inside the patient's body.

14 To help us understand what radioactive emissions are we use this model of an atom:

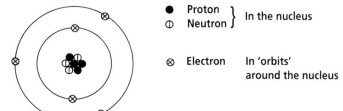

15 The number of electrons in an atom must match the number of protons. Every element has its own characteristic proton (and electron) number. But you can have atoms of the same element with different numbers of neutrons. These are called different **isotopes** of the element.

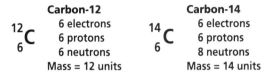

16 α, β and γ radiations come from the nuclei of the atoms of radioactive substances. If the nucleus is unstable, it may emit an α particle, a β particle or a γ ray.

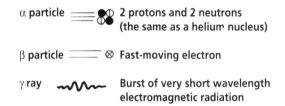

17 Waves from many parts of the electromagnetic spectrum are used in medical diagnosis:

Radio waves	Used in MRI (magnetic resonance imaging)
Infra-red radiation	Thermal imaging shows hot regions
X-rays	Shadow pictures of bones and internal organs; CT scans
γ rays	Gamma emitting tracers to investigate different organs; gamma camera

18 Sound waves can be used for 'seeing'. By timing the interval between a sound pulse and its echo you can work out how far away an object is. This is used by bats and in echo sounders.

19 For seeing inside the body, you get better pictures if you use sound waves of short wavelength (**ultrasound**). These waves are too high pitched for humans to hear. A detector picks up the echoes and produces a picture. Ultrasound does not cause ionization, so it is safer than X-rays.

In brief

Sound Reproduction

1 All sounds are produced by something vibrating. Vibrations travel from the source through a medium (usually air) as pressure pulses – **compressions** and **rarefactions.** These pulses can then set other things vibrating.

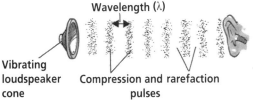

Wavelength (λ)

Vibrating loudspeaker cone

Compression and rarefaction pulses

Air pulses make the eardrum vibrate - we hear sound

2 Sound travels as a **longitudinal wave.** This means that the medium vibrates back and forth in the same direction as the sound is travelling.

3 The number of vibrations in a second is called the **frequency** of a sound. It is measured in **hertz** (Hz). A frequency of 300 Hz means 300 vibrations in a second.

4 The distance between one compression pulse and the next is called the **wavelength** (λ) of the sound. There is a simple connection between frequency, wavelength and the speed at which the sound travels:

$$\text{speed} = \text{frequency} \times \text{wavelength}$$

$$v = f\lambda$$

5 An **oscilloscope trace** is a graph showing how the air pressure changes in a sound wave. Loud sounds have a large amplitude of vibration; high-pitched sounds have a high frequency.

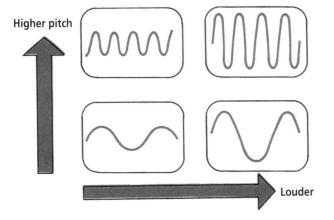

Higher pitch

Louder

6 **Music** is a mixture of sounds of many different frequencies. Even the individual notes are mixtures of several related frequencies. The mixture is different for each musical instrument. This is one of the things which helps us tell instruments apart.

7
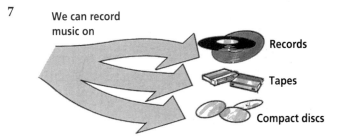

We can record music on

Records

Tapes

Compact discs

Making a recording: A recording is a copy of the original sound pattern. For a good recording, the microphone has to pick up all the frequencies in the music and the amplifier has to amplify them all equally.

Listening to recorded sound: A recording sounds good if your sound system can reproduce all the frequencies which were in the original sound in the same mixture.

8 Sound waves are **diffracted** as they pass through small gaps. Sound waves from two sources can **interfere** with each other. These wave properties are important when designing loudspeakers and setting up sound systems.

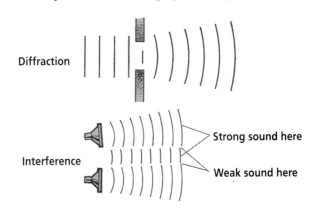

Diffraction

Interference

Strong sound here

Weak sound here

9 Every object has a **natural frequency** at which it can vibrate. If a wave of this exact frequency strikes the object, it can set it vibrating strongly. This is called **resonance.**

▶

10 Sound recording and reproduction involve electromagnetic devices. These all use the magnetic effects of electricity:
   • loudspeakers
   • microphones
   • tape recorders
   • record player pickups
   • motors in turntables, tape recorders and CD players.

11 There is a **magnetic field** – a region where you can detect magnetic effects – around any wire which is carrying an electric current. A much stronger field can be produced by winding a coil of wire round an iron core. This makes an **electromagnet.**

An electromagnet in action

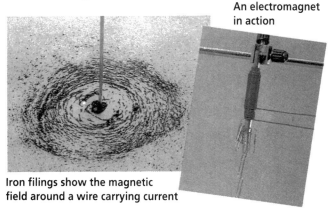

Iron filings show the magnetic field around a wire carrying current

12 If you pass an electric current through a wire in a magnetic field, a force makes the wire move sideways.

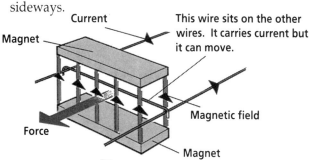

Current

Magnet

This wire sits on the other wires. It carries current but it can move.

Magnetic field

Force

Magnet

A motor uses this force to produce continuous rotation.

13

If you push a magnet into a coil of wire...

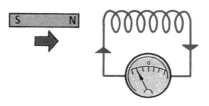

...there is a brief pulse of current in the wire while the magnet is moving.

If you take the magnet out of the coil again...

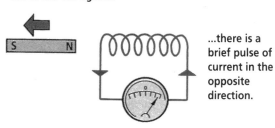

...there is a brief pulse of current in the opposite direction.

The same thing happens if you keep the magnet still and move the coil. A small current is generated while the magnet is moving relative to the coil. This is called **electromagnetic induction.**

14 Electromagnetism and electromagnetic induction are opposite effects.

Electromagnetism	Electromagnetic induction
What you do: *Pass a current through a wire in a magnetic field*	What you do: *move a wire in a magnetic field*
The result... the wire *moves*.	The result... a *current* is generated in the wire.

15 Information can be stored and transmitted in **analogue** or **digital** form. An analogue signal varies smoothly and can take any value. Digital information is a series of whole numbers. It is like taking a series of readings from a graph of the signal. The information can be stored and transmitted simply as this list of numbers. Information can be transmitted more accurately in digital form.

In brief

Sports Science

1 Most sports involve using your muscles to do physical **work**. Scientists use a very precise definition of work. Work is done when a force makes an object move.

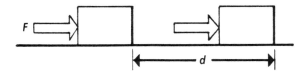

The amount of work is calculated from the equation:

work = force × distance
$$w = Fd$$

The distance must be measured in the direction of the force.

2 The units of work are **newton metres**, which are also called **joules (J)**. One joule is the amount of work done by a force of one newton when it moves an object a distance of one metre.

3 When a force does work, it transfers **energy** from one form to another. The amount of work done is equal to the amount of energy transferred. Like work, energy is also measured in joules.

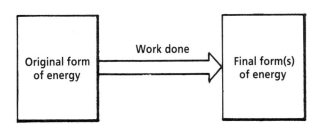

4 **Power** is the rate at which a person or a machine can do work:

$$\text{power} = \frac{\text{work done}}{\text{time taken}}$$

Power can also be defined as the rate at which energy is transferred in the process:

$$\text{power} = \frac{\text{energy transferred}}{\text{time taken}}$$

5 Power is measured in **watts (W)**. One watt is one joule per second. Larger powers are measured in kilowatts (kW) or megawatts (MW). 1 kilowatt is 1000 watts; 1 megawatt is 1 000 000 watts.

6 Fitness tests may involve measuring:
- the power of certain muscles
- the time you take to recover after exercise.

7 When you exercise, a chemical reaction takes place in your muscles. Glucose and oxygen are used up and carbon dioxide, water and lactic acid are produced. Work is done and heat is also produced.

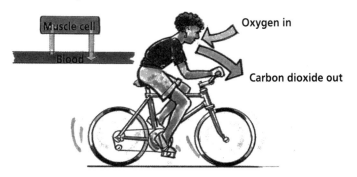

8 When you lift an object, you increase its **gravitational potential energy**.

The gain in gravitational potential energy is calculated by the equation:

$$E_p = mgh$$

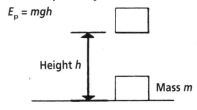

9 Any moving object has **kinetic energy**.

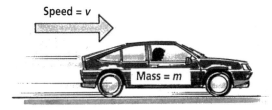

The amount of kinetic energy depends on the mass of the object and its speed:

$$E_k = \tfrac{1}{2}mv^2$$

▶

10 If an object slides or rolls down a slope, it loses gravitational potential energy and gains kinetic energy. If there is no friction, the amount of kinetic energy it gains is *equal* to the amount of gravitational potential energy it has lost.

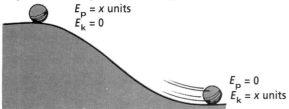

This means that the ball's speed at the bottom of the hill does not depend on the shape of the slope, but only on the drop in height.

The same is true in the other direction. If an object slides or rolls up a slope, it loses kinetic energy and gains gravitational potential energy. If there is no friction, the amount of gravitational potential energy it gains is equal to the amount of kinetic energy it loses.

11 **Muscles** exert forces by contracting. In vertebrates, muscles are attached to the bones of the skeleton and operate joints like a system of levers.

Most joints are moved by a pair of muscles, one moving the joint in one direction and the other moving it back.

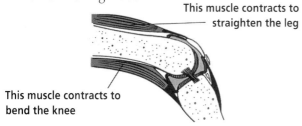

12 The **turning effect** of a force is calculated by the equation:

turning effect = force × distance from pivot

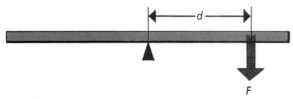

The distance is measured *at right angles* to the direction of the force. (Notice how this is different from the equation for work in *In Brief 1*.)

13 An object **balances** if the total turning effect in the clockwise direction is equal to the total turning effect in the anticlockwise direction.

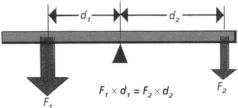

14 You can use **levers** to move something more easily. Some levers multiply the *force* you apply. The force you apply is smaller than the load. But the load moves a shorter distance than the applied force.

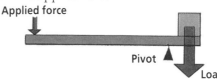

Using a teaspoon to open a tin of syrup is an example of this kind of lever.

Other levers are used to multiply the *distance* the applied force moves. The force you apply moves a short distance and makes the load move a longer distance. But the applied force has to be bigger than the load.

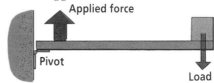

Lifting a weight in your hand with your forearm outstretched is an example of this kind of lever.

15 Every object behaves as if all its mass were concentrated at a single point called the **centre of mass**. When an object hangs, its centre of mass is always directly below the point of suspension.

16 An object is **stable** if its centre of mass is low, or if it has a wide base. You have to tip it a long way to push its centre of mass past the edge it is pivoting on.

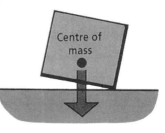

In brief

The Atmosphere

1 The **atmosphere** is a thin layer of gases around the Earth. The inner layer is the **troposphere.** The gases in the troposphere get gradually thinner as you go higher. At about 20 km up, almost all the gas is gone. Above this is the **ionosphere** – the outer layer of the atmosphere which contains ions (charged atoms).

2 The pie chart shows the gases which make up the atmosphere.

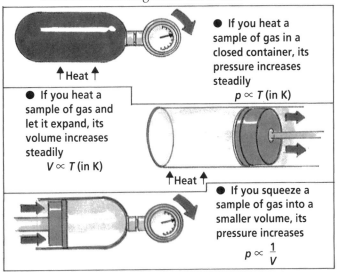

Nitrogen, 79%

Oxygen, 20%

Argon, 1%

Other gases (including carbon dioxide, 0.03%)

3 The atmosphere is so large that the air temperature and pressure vary within it. There are different **air masses** in different places, for example warm, humid air near the equator and colder, drier air near the poles.

4 These air masses are continually moving above the Earth. This movement is caused by the heating effect of the Sun's radiation. Air near the equator is heated strongly and becomes less dense. So it rises and colder air moves in to take its place. This causes **convection currents.**

5 To understand how the atmosphere behaves it helps to know how gases behave in simpler situations. Investigations show that:

● If you heat a sample of gas in a closed container, its pressure increases steadily
$p \propto T$ (in K)

↑ Heat ↑

● If you heat a sample of gas and let it expand, its volume increases steadily
$V \propto T$ (in K)

↑ Heat ↑

● If you squeeze a sample of gas into a smaller volume, its pressure increases
$p \propto \dfrac{1}{V}$

6 The behaviour of gases can be explained using the **kinetic model of gases.** A gas consists of lots of tiny particles moving rapidly in all directions. Heating the gas makes the particles speed up. Pressure is caused by collisions of the particles with the sides of the container.

7 To convert temperatures from degrees Celsius to kelvin, add 273. So 0°C is 273 K and 100°C is 373 K.

8 The air near the equator is less dense than in other places – the pressure is low. The air near the poles is colder and denser – the pressure is high. Air moves from high pressure areas to low pressure areas, causing winds. As the Earth is rotating, the winds tend to swirl around the high and low pressure regions.

9 Sometimes a cold air mass meets a warm air mass, forming a **weather front.** The denser cold air forces the warmer air upwards. The warmer air cools. Water vapour in the rising warmer air begins to condense into tiny water droplets, forming clouds. The droplets join together and fall as rain. Rain is associated with weather fronts.

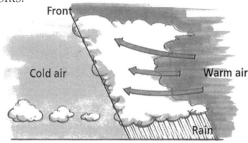

Front

Cold air

Warm air

Rain

10 **Isobars** on a weather map join points where the air pressure is the same. They help us to locate weather fronts and to predict the wind direction.

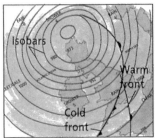

Isobars

Warm front

Cold front

11 Energy is involved when a substance changes from one **state** to another. When ice melts to form water, it absorbs energy to break the strong bonds holding the water molecules together as solid ice. When liquid water cools to form snow crystals, it releases energy.

In brief

The Earth in Space

1 The **solar system** consists of nine **planets**, and a belt of smaller bodies called **asteroids**, orbiting the Sun. **Comets** also orbit the Sun and are part of the solar system.

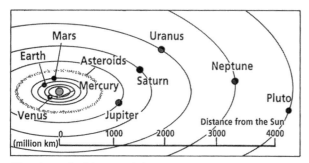

The planets' orbits round the Sun

The orbits of the planets are ellipses which are almost circular. Some planets have their own satellites or **moons** which orbit the planet.

2 We can see some patterns in the solar system:
   ● all the planets orbit in the same plane
   ● the surface temperatures of the planets get lower the further they are from the Sun
   ● the larger planets (Jupiter, Saturn, Uranus and Neptune) are made mainly of substances with low melting points
   ● the smaller planets are rocky.

3 These observations, along with others, suggest that the solar system formed from a single **nebula**, an enormous swirling cloud of dust and gas. The gravitational attraction between the particles of matter pulled them together to form the Sun and the planets.

4 The Sun is a **star**. Its fuel is hydrogen. In a process called **nuclear fusion**, hydrogen nuclei join together to make helium nuclei, releasing energy. All stars are powered by nuclear fusion.

5 When a star's fuel runs out, the star may collapse. The gravitational attraction between different parts of the star's matter pulls them all closer and closer together. The matter contracts to a very small dense centre, which may become a **white dwarf star**, a **neutron star** or a **black hole**.

6 The first stage of **space travel** is lift-off from the Earth. Because of the gravitational pull of the Earth, very large forces are needed to accelerate a rocket upwards. It has to reach a speed of 11 km/s or it will eventually fall back to Earth.

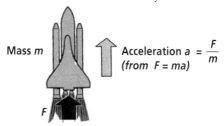

Mass m

Acceleration $a = \dfrac{F}{m}$
(from $F = ma$)

F

7 A spacecraft in deep space, far away from all planets or stars, moves at a steady speed in a straight line when its rocket motors are turned off. This is an example of Newton's first law of motion. A spacecraft in Earth orbit also continues to move at a steady speed without any rocket force.

8 In order to move, an object needs something to push back on. A rocket in space pushes on its own fuel. The backwards push of the rocket on the exhaust gases causes an equal and opposite forwards push on the rocket itself.

Force exerted by rocket on exhaust gases

Force exerted by exhaust gases on rocket

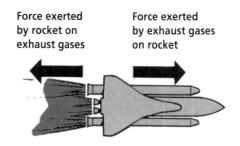

9 An object thrown vertically upwards slows down steadily until it reaches the top of its motion.
On the way down, it speeds up again.

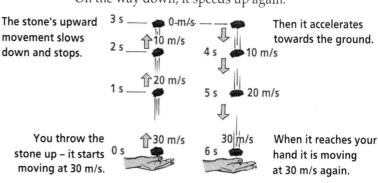

The stone's upward movement slows down and stops.

3 s — 0 m/s

2 s — ⇧10 m/s

1 s — ⇧20 m/s

⇧30 m/s

You throw the stone up – it starts moving at 30 m/s.

0 s

Then it accelerates towards the ground.

4 s 10 m/s

5 s 20 m/s

30 m/s

6 s

When it reaches your hand it is moving at 30 m/s again.

10 An object projected forwards falls downwards at exactly the same rate as a free-falling object, provided air resistance can be ignored. The forwards motion makes no difference to how fast it drops.

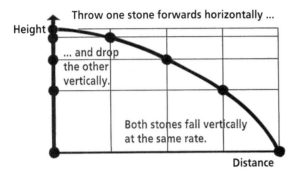

11 Because the Earth is a sphere, an object projected forwards at a high speed may fall towards the Earth at exactly the same rate as the Earth's surface curves away below it! It will continue moving round the Earth, falling towards it all the time, but never getting there. It has become a satellite in orbit.

12 An object can move in a circle only if there is a force pulling it towards the centre. Without such a force it will fly off in a straight line in the direction it was travelling when the force stopped acting on it.

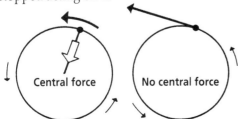

13 The force which keeps the Moon and artificial satellites in orbit around the Earth is the Earth's **gravitational force**. The Sun's gravitational force keeps the planets in their orbits around the Sun.

14 Every object exerts a gravitational force on every other object. This force is always an attraction. The greater the mass of the objects, the bigger the gravitational force. The gravitational force gets smaller as the distance between the two objects increases, but it never becomes zero.

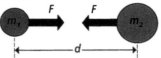

F increases as m_1 and m_2 increase.
F decreases as *d* increases.

15 The **weight** of an object (in newtons) is the size of the gravitational force on the object. This depends on where the object is. The **mass** of an object (in kilograms) is a measure of how much force is needed to make the object accelerate – its **inertia**. The mass of an object never changes, no matter where the object is. But the object's weight changes if you take it up a mountain or into space because the gravitational force on it changes.

16 At the Earth's surface, the strength of the gravitational field is roughly 10 N/kg. So a mass of 1 kg has a weight of about 10 N.
On the Moon, the gravitational field strength is 1.6 N/kg, so a kilogram will weigh 1.6 N. But it would be just as difficult to make the kilogram speed up on the Moon as on Earth.

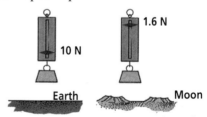

17 Our knowledge of the Universe is based on **observation**. Scientists develop **theories** to explain what they observe. Sometimes these theories can be used to make exact **predictions**. If **new observations** agree with the predictions, this makes us more confident in the theory.
Other advances in astronomy have come from new observations which could *not* be explained by the theories available at the time. This led to the development of new and more powerful theories.

In brief

Transporting Chemicals

1 The chemical industry is a large and important part of the manufacturing industry. Sometimes it converts raw materials into products which we buy, such as oil into petrol. But often it first converts the raw materials into chemicals (intermediates) which are then used to make products.

2 The factories used to make products are built at different places.

The choice of site is influenced by:

● source of raw materials
● road, rail and sea links
● available workforce
● environmental factors
● energy supply
● water supply.

3 The intermediate or bulk chemicals can be transported to other factories by:

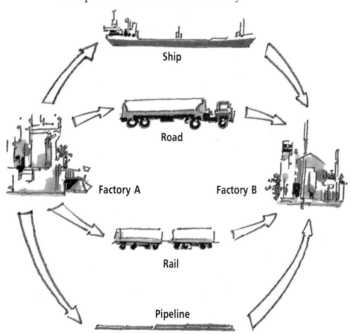

The method used depends on geographical, economic, social and environmental factors.

4 Different chemicals often have different properties. For example, some burn, some are corrosive and some are gases. When they are transported around the country their containers must be carefully labelled. The

Hazchem code system of labelling tells the police and fire services dealing with an emergency what sort of chemicals are involved.

5 An **element** is a substance which cannot be broken down into anything simpler. The smallest particle of an element is called an **atom.** Atoms of one element are different from atoms of all other elements. Each element is represented by a **symbol**, which is shorthand for the element.

Compounds contain the atoms of more than one element joined together (not just mixed up). Each pure compound is represented by a formula which is shorthand for the compound.

For example,

$$CaCO_3$$

represents calcium carbonate. This tells you that it contains

calcium, carbon and oxygen

and that these elements are present in the ratio

1 : 1 : 3

The smallest particles of some compounds and elements are called **molecules**. For example, H_2O represents a molecule of water and H_2 a molecule of hydrogen.

6 Chemical reactions between elements or compounds can be represented by word equations. They list the starting substances (reactants) and the products. For example, the burning of natural gas (methane) can be represented by:

methane + oxygen → carbon dioxide + water

Alternatively, an equation which is made up of symbols and formulas can be used:

$$CH_4 + 2O_2 \rightarrow CO_2 + 2H_2O$$

This is called a **balanced** equation because it has the same number of atoms of each element on each side.

7 The **periodic table** is a way of arranging and displaying the elements which helps you remember the similarities and differences between them. It collects similar elements together in vertical columns called **groups**. The horizontal rows are called **periods**.

In brief

Waste Not Want Not

1 Every household produces waste.

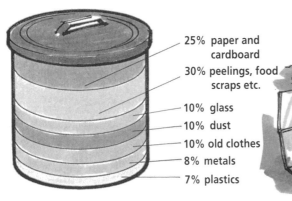

- 25% paper and cardboard
- 30% peelings, food scraps etc.
- 10% glass
- 10% dust
- 10% old clothes
- 8% metals
- 7% plastics

2 We can either dump or burn our rubbish. Useful materials can be reclaimed from rubbish and then reused or recycled.

3 Reusing and recycling materials conserves resources and can save energy and reduce pollution.

4 We need to dispose of waste carefully to avoid pollution.

5 Some waste materials can be decomposed by micro-organisms. These materials are **biodegradable.** The resources in biodegradable waste are recycled naturally.

6 Waste which cannot be broken down by micro-organisms is described as **non-biodegradable.** Some non-biodegradable materials such as aluminium can be reclaimed by physical or chemical methods. Others cannot and so we cannot use them again.

7 Sewage is mostly water containing suspended and dissolved materials. Untreated sewage is a powerful pollutant.

Raw sewage entering the sea.

8 Sewage is treated at a sewage works. The treatment uses both physical and biological processes.

9 Micro-organisms play an important role in sewage treatment. They decompose sewage into simple chemicals.

10 Sewage treatment purifies water and releases useful chemicals for reuse. It is an important recycling system.

▶

11 Nitrogen is an essential element for life. There is a natural system which recycles nitrogen. Soil micro-organisms play an important role in this process. Human activity affects the nitrogen cycle.

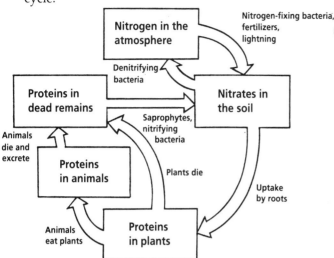

12 Nitrogen-fixing bacteria convert atmospheric nitrogen into nitrates. **Nitrifying bacteria** convert ammonium compounds released in decomposition into nitrates. These nitrates can be used by plants. **Denitrifying bacteria** convert nitrates into nitrogen.

- **Nitrogen-fixing bacteria**:
  nitrogen → nitrates

- **Nitrifying bacteria**:
  ammonium compounds → nitrites → nitrates

- **Denitrifying bacteria**:
  nitrates → nitrogen

13 Crops remove nitrogen from the soil. Farmers use fertilizers or manure to replace the nitrogen and keep the soil fertile.

14 All living things contain carbon. There is a natural system which recycles carbon.

Respiration and decomposition release carbon dioxide into the atmosphere. Photosynthesis removes carbon dioxide from the atmosphere. Human activity affects the carbon cycle.

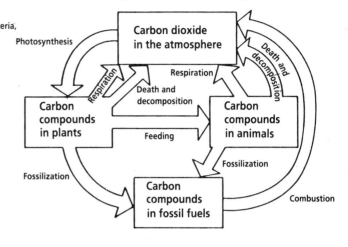

15 The burning of fossil fuels releases carbon dioxide. Industrialization has increased the burning of fossil fuels and so increased the concentration of carbon dioxide in the atmosphere.

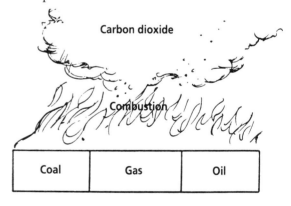

16 Deforestation has reduced the number of trees. Trees absorb carbon dioxide by photosynthesis. So deforestation has contributed to the increased concentration of carbon dioxide in the atmosphere.

17 No new matter is ever created in the world. All the existing matter is rearranged many times to form new substances and new organisms.

Heinemann Educational, Halley Court, Jordan Hill, Oxford OX2 8EJ

© University of York Science Education Group 1992

First published 1992   93 94 95 96 12 11 10 9 8 7 6 5 4
Design by KAG Design Ltd
ISBN 0 435 63028 8